T0340201

Industrial Piping and Equipment Estimating Manual

Industrial Piping and Equipment Estimating Manual

Kenneth Storm

Gulf Professional Publishing
An imprint of Elsevier

Gulf Professional Publishing is an imprint of Elsevier
50 Hampshire Street, 5th Floor, Cambridge, MA 02139, United States
The Boulevard, Langford Lane, Kidlington, Oxford, OX5 1GB, United Kingdom

Notices
Knowledge and best practice in this field are constantly changing. As new research and
experience broaden our understanding, changes in research methods, professional practices,
or medical treatment may become necessary.

Practitioners and researchers must always rely on their own experience and knowledge in
evaluating and using any information, methods, compounds, or experiments described herein.
In using such information or methods they should be mindful of their own safety and the
safety of others, including parties for whom they have a professional responsibility.

To the fullest extent of the law, neither the Publisher nor the authors, contributors, or editors,
assume any liability for any injury and/or damage to persons or property as a matter of
products liability, negligence or otherwise, or from any use or operation of any methods,
products, instructions, or ideas contained in the material herein.

Library of Congress Cataloging-in-Publication Data
A catalog record for this book is available from the Library of Congress

British Library Cataloguing-in-Publication Data
A catalogue record for this book is available from the British Library

ISBN: 978-0-12-813946-2

For information on all Gulf Professional Publishing publications visit
our website at https://www.elsevier.com/books-and-journals

Working together
to grow libraries in
developing countries

www.elsevier.com • www.bookaid.org

Publishing Director: Joseph P. Hayton
Senior Acquisition Editor: Katie Hammon
Editorial Project Manager: Katie Chan
Production Project Manager: Mohanapriyan Rajendran
Designer: Mark Rogers

Typeset by TNQ Books and Journals

Contents

Preface

This edition of "Industrial Piping and Equipment Estimating Manual" provides the principles and techniques for estimating process piping and equipment. It is not a manual about estimating only. Estimating cannot precede without labor productivity and analysis. The manual begins with the introduction devoted to labor, productivity measurement, collection of historical data, estimating methods, and factors affecting construction labor productivity and impacts of overtime. The direct craft man hours in this manual were determined from time and methods for the measurement of construction labor for field erection of process piping and equipment for industrial projects located throughout the country.

Then eight sections provide the estimate tables for process piping and equipment scopes of work, estimate man-hour tables and estimate sheets, and sample estimates and statistical applications. The purpose of this manual is to provide a comprehensive and accurate method for compiling piping and equipment direct craft man-hour estimates for the following industrial facilities:

Combined Cycle Power Plant
Simple Cycle Power Plant
Refinery and Hydrogen Plant
Compressor Station
Biomass Plant
Ethanol Plant

This manual is intended for project managers, estimators, engineers and project field staff to prepare direct craft man-hour estimates for budgets, RFP's, bid proposals, and field change orders. The principles and techniques for estimating process piping and equipment will allow the estimator to accurately determine the actual direct craft man hours, based on historical and quantitative data, for the complete field installation of mechanical equipment and piping for a given industrial facility.

The estimating methods in this manual will allow the experienced estimator to use the comparison method to estimate similarities and differences between proposed and previously installed equipment and the unit quantity method will be a final check on actual man hours compared to estimate man hours. The manual does not include cost and man hours for material, equipment usage, indirect craft and supervision, project staff, warehousing and storage, shop fabrication, overheads, and fee. The direct craft man-hour estimate is the basis for the estimator to obtain the project schedule and the man hours and cost for indirect craft and supervision, project staff, construction equipment, material, subcontractors, mobilize and demobilize, site general conditions, overhead and fee. In addition, the estimator must determine all factors that will affect direct craft labor productivity and overtime impacts. Review of the Preface and Introduction will enable the estimator to understand labor productivity, productivity measurement, collection of historical data, estimating methods, and labor factors and loss due to labor productivity and overtime impacts. The sample estimates, in Section 8, will illustrate how to apply the estimate tables and sheets to prepare detailed direct craft estimates and the statistical applications to construction will provide the statistical methods to forecast man-hour analysis by graphical and analytical techniques.

To apply the principles and techniques, estimate man-hour tables and sheets, the estimator must be familiar with the introduction on the following pages and "Section 8: Sample Estimates and Statistical Applications to Construction."

Introduction

Introduction

The estimate data in this manual has been verified by measurement, project cost reports from field erection of process piping and equipment, one-cycle time studies, and the data are revised continuously due to construction design, engineering, labor skill, material, equipment, and procedures.

Information for man hour analysis is obtained from the foreman's report. The foreman's report is used to find the number of man hours for a task. The report is used for cost and time control. The cost engineers and welding quality control in the field monitor and verify the work.

These reports are collected for the field installed piping and equipment. From the reports and review of the specifications, codes, and drawings the cost engineer and estimator will examine the data for consistency, completeness, and accuracy. Reports are collected for similar work and the data are entered into a spreadsheet. The spreadsheet prepares the data for mathematical analysis. The engineer and estimator determine the productivity rate. The rate is used for future cost analysis and estimating similar scopes of work.

The estimate data are based on "standard," which is defined as "forming a basis for comparison." The standard unit man hour involves these considerations: the work is being performed by a contractor who is familiar with all conditions at the job site; the project has the proper supervision; the workers are familiar with and skilled in performance of the work task; and there is an adequate supply of labor. There are clarifications and exceptions stated for the application of the data.

Labor Productivity and Analysis

Labor

Labor productivity is concerned with direct craft labor. Direct craft labor time means the craft is working on the field erection of process piping and equipment. The "man hour" is dependent on the historical value of time spent erecting pipe and equipment on industrial projects.

This basic unit is defined as one worker working for 1 h.

Examples of man hour units:

HRSG—Seal Weld Side, Roof and Floor Casing Field Seams 0.35 MH/LF
Welding Butt Weld, Carbon Steel, Arc-Uphill WT <= 0.375″ 0.50 MH/Diameter Inch

Productivity Measurement

Historical records provided the direct craft man hour data for field installation of piping and equipment. Two methods for the measurement of construction time were used to collect, analyze, and compile the actual man hour data in this manual.

Nonrepetitive one-cycle time t study
Foreman report—job cost-by-cost code and type

Nonrepetitive One-Cycle Time Study and Man Hour Analysis

Nonrepetitive time study was used for construction of direct craft long-cycle scopes of work. The time study provides man hour information for cost analysis and estimating. The study requires continuous timing with an electric timer or video camera. A camcorder with video playback is used, and video tape can be returned to determine acceptable methods and the time required for the work.

Once the time study is complete, the craft foreman determines the operation or task and calculates a time for it. The calculation is the net time less any reductions for task unrelated to the timed task. Normal time is found by multiplying a selected time for the task or cycle by the rating factor:

$$T_n = T_o \times RF$$

where T_n=normal time, hours; T_o=observed time, hours; RF=rating factor, arbitrarily set, number.

Example: If the craft worker is fast, then the RF>1.0, say example as 110%.

Task time is 1.8 h, then normal time is $(1.8 \times 1.10) = 1.98$ h.

If the worker is rated 90% then RF<1 and normal time is $(0.90 \times 1.8) = 1.62$ h.

The rating factor allows the "sample" observation to be adjusted for normal workers to arrive at a true value.

Normal time does not include factors that affect labor productivity. Allowances for these factors are divided into three components: personal, fatigue, and delay (PF&D).

Process of timing the cycle:

Idle time is excluded; craft takes breaks for coffee and rest room; allowance for personal is 5%.

Fatigue is physiological reduction in ability to do work; allowance for fatigue is 5%.

Delays beyond the worker's ability to prevent; allowance for delays is 5%.

Productivity time in the work day is inversely proportional to the amount of PF&D allowance; the allowance is expressed as a percent of the total work day.

PF&D allowance is generally in the range of 10%–20%. Allowance Multiplier is expressed as follows:

$$F_a = 100\% / (100\% - PF\&D\%)$$

where F_a=allowance multiplier for PF&D, number; PF&D=personal, fatigue, and delay allowance, percentage; standard productivity—time required by a trained and motivated worker or workers to perform construction task while working at normal tempo.

$$H_s = T_n \times F_a$$

where H_s=standard time for a construction task per unit of effort, hour.

The allowance for PF&D is 15%, which becomes an allowance multiplier of 1.176.

The estimate data in this manual is based on "standard," which is defined as "forming a basis for comparison." The standard unit man hour involves these considerations: the work is being performed by a contractor who is familiar with all conditions at the job site; the project has the proper supervision; the workers are familiar with and skilled in performance of the work task; and there is an adequate supply of labor. There are clarifications and exceptions stated for the application of the data.

Code of Accounts and Foreman's Report—Job Cost-by-Cost Code and Type

The estimator provides the cost engineer the estimated budget direct craft man hours and cost for the project to be erected in the field. The cost engineer uses the code of accounts to set up the tracking system for the project erection (Table 1).

Table 1 **Illustration of Portion of the Code of Accounts for Erection of an HRSG**

Cost	Phase Code Description	Craft	Budget MH
Code	**UNIT ONE**		
xxxxxx	HRSG—Casing columns	BM	4838
xxxxxx	HRSG—Duct work SCR and Inlet	BM	2183
xxxxxx	Erect—Gas baffles	BM	1620
xxxxxx	HRSG—Modules	BM	880
xxxxxx	HRSG—Pressure vessels	BM	420
xxxxxx	HRSG—Steel (ladders, platforms, and grating)	BM	2211
xxxxxx	HRSG—Stack and breeching	BM	2582
xxxxxx	HRSG—Skids	BM	80
xxxxxx	HRSG—SCR and CO_2 internals	BM	708
xxxxxx	HRSG—Receive, off load, and Haul to Hook	BM	978
	Direct Craft Man Hours		**16,500**

Foreman's Report

Information for man hour analysis is obtained from the foreman's report. The foreman's report is used to find the number of man hours for a task. The report is used for cost and time control. The cost engineers and welding quality control in the field monitor and verify the work.

These reports are collected for the field-installed piping and equipment. From the reports and review of the specifications, codes, and drawings the cost engineer and estimator will examine the data for consistency, completeness, and accuracy. Reports are collected for similar work and the data are entered into a spreadsheet. The spreadsheet prepares the data for mathematical analysis. The engineer and estimator determine the productivity rate. The rate is used for future cost analysis and estimating similar scopes of work (Tables 2–4).

Table 2 **Illustration of Portion of Foreman's Report for Erection of an HRSG**

Cost	Phase Code Description	W/E	Last Name	HR Type	Hours
Code	UNIT ONE				
xxxxxx	HRSG—Casing columns	Date	Smith	ST	8
xxxxxx	HRSG—Duct work SCR and inlet	Date	Smith	ST	4
xxxxxx	Erect—Gas baffles	Date	Smith	ST	8
xxxxxx	HRSG—Modules	Date	Smith	ST	8
xxxxxx	HRSG—Pressure vessels	Date	Smith	ST	8
xxxxxx	HRSG—Steel (ladders, platforms and grating)	Date	Smith	ST	4
xxxxxx	HRSG—Stack and breeching	Date	Smith	ST	6
xxxxxx	HRSG—Skids	Date	Smith	ST	4
xxxxxx	HRSG—SCR and CO_2 Internals	Date	Smith	ST	8
xxxxxx	HRSG—Receive, off Load and haul to hook	Date	Smith	ST	4

Table 3 Illustration of Portion of Job Cost-by-Cost Code and Type for Erection of an HRSG

Cost	Phase Code Description	Craft	Actual MH
Code	UNIT ONE		
xxxxxx	HRSG—Casing columns	BM	4648
xxxxxx	HRSG—Duct work SCR and inlet	BM	1053
xxxxxx	Erect—Gas baffles	BM	831
xxxxxx	HRSG—Modules	BM	535
xxxxxx	HRSG—Pressure vessels	BM	400
xxxxxx	HRSG—Steel (ladders, platforms and grating)	BM	1716
xxxxxx	HRSG—Stack and breeching	BM	2096
xxxxxx	HRSG—Skids	BM	88
xxxxxx	HRSG—SCR and CO_2 internals	BM	308
xxxxxx	HRSG—Receive, off load and haul to hook	BM	1143
	Direct Craft Man Hours		**12,818**

Table 4 Illustration of Portion of Tracking Report for Erection of an HRSG

Cost	Phase Code Description	Budget	Actual	Percent	Productivity
Code	UNIT ONE	MH	MH	Complete	(A/E)
xxxxxx	HRSG—Casing columns	4838	4648	100.00%	1.04
xxxxxx	HRSG—Duct work SCR and inlet	2183	1053	100.00%	2.07
xxxxxx	Erect—Gas baffles	1620	831	100.00%	1.95
xxxxxx	HRSG—Modules	880	535	100.00%	1.65
xxxxxx	HRSG—Pressure vessels	420	400	100.00%	1.05
xxxxxx	HRSG—Platforms	2211	1716	100.00%	1.29
xxxxxx	HRSG—Stack and breeching	2582	2096	100.00%	1.23
xxxxxx	HRSG—Skids	80	88	100.00%	0.91
xxxxxx	HRSG—SCR and CO_2 internals	708	308	100.00%	2.30
xxxxxx	HRSG—Receive, off load and haul	978	1143	100.00%	0.86
	Direct Craft Man Hours	**16,500**	**12,818**	**100.00%**	**1.29**

Estimating Methods

Comparison Method

The comparison logic is based on estimating similarities and differences for proposed equipment installation and previously installed equipment for which historical man hour data are available.

Designate MHc = historical man hour (know man hour estimate); MH_a = Equipment quantity is increased; MH_β = Equipment quantities are decreased.

$MHc <= MH_a$; MH_a is directly proportional to MHc; MH_β is a lower bound and logic is expanded to $MH_\beta <= MHc <= MH_a$.

The estimate that is either above or below the known estimate must satisfy the work scope and quantity take off. The less difference between the equipment estimate and the original comparison estimate the better the comparison.

The comparison method provides the opportunity to compare the proposed estimate to the previous estimate. The estimate data must be current and need to be verified by measurement, project cost reports, one-cycle repetitive time study, and historical experience, and the data need to be revised continuously with the many combinations of construction design and engineering, labor skill, material, equipment, and procedures. The comparison method in this manual will be more effective if maintained and revised by the contractor, engineer, and owner.

Estimating Equipment by Comparison

Determine scope of work for the proposed equipment installation and make a detailed quantity takeoff.

The quantity takeoff is either above or below the known quantity and satisfies the scope.

The lower or upper bound is the reference to increased or decreased quantities.

The known quantity is historical data from previous field installations.

The Main Advantages of Comparison Method

1. Estimate is based on comparison with actual unit man hours and quantities.
2. The scope and quantity differences can be identified and the impacts estimated.
3. Comparison method applies to any complexity of design, bid, and contract for a project.

A Simple Example:

Estimator is using comparison to estimate an air filter house to be erected in a combined cycle power plant.

Example—Comparison Method. F Class CTG Installation Estimate; Air Filter House

$$MH_\beta <= MHc <= MH_a$$

where MHc = historical man hour; MH_a = equipment quantity is increased; MH_β = equipment quantities are decreased (Tables 5–7).

Comparison of Estimated Direct Craft Man Hours	Man Hours
MH_β = Equipment quantities are decreased; estimated man hours are decreased	1557
MHc = known unit man hour based on historical data	1730
MH_a = Equipment quantity is increased; estimated man hours are increased	1826

Table 5 MHc = Historical Man Hour Estimate for Inlet Filter House

Estimate—Inlet Filter House	Historical			Estimate				
Description	MH	Quantity	Unit	Quantity	Unit	BM	IW	MW
Air Filter System								
Inlet filter house						**1730**	**0**	**0**
Sloped roof	8.40	2.1	Ton	2.1	Ton	18		
Hopper	20.00	1.0	Ton	1.0	Ton	20		
Pipe/tubing	1.94	49.7	LF	49.7	LF	96		
Hand rails	0.25	21.1	LF	21.1	LF	5		
Filter cartridges	0.75	1020.0	EA	1020.0	EA	765		
Filter modules (center)	8.40	15.5	Ton	15.5	Ton	130		
Filter modules (left and right hand)	8.40	9.3	Ton	9.3	Ton	78		
Walkway assembly (center)	8.40	4.7	Ton	4.7	Ton	39		
Walkway assembly (LH and RH)	8.40	3.5	Ton	3.5	Ton	29		
D15 weather hoods	8.40	9.9	Ton	9.9	Ton	83		
Field joints	0.75	620.6	LF	620.6	LF	465		

Table 6 MH_a = Estimate for Increased Inlet Filter House Quantities

Estimate—Inlet Filter House		Historical				Estimate		
DESCRIPTION	MH	Quantity	Unit	Quantity	Unit	BM	IW	MW
Air Filter System								
Inlet filter house						**1826**	**0**	**0**
Sloped roof	8.40	2.4	Ton	2.4	Ton	20		
Hopper	20.00	1.1	Ton	1.1	Ton	22		
Pipe/tubing	1.94	54.7	LF	54.7	LF	106		
Hand rails	0.25	23.2	LF	23.2	LF	6		
Filter cartridges	0.75	1020.0	EA	1020.0	EA	765		
Filter modules (Center)	8.40	17.1	Ton	17.1	Ton	143		
Filter modules (left and right Hand)	8.40	10.2	Ton	10.2	Ton	85		
Walkway assembly (center)	8.40	5.1	Ton	5.1	Ton	43		
Walkway assembly (LH and RH}	8.40	3.8	Ton	3.8	Ton	32		
D15 weather hoods	8.40	10.9	Ton	10.9	Ton	91		
Field joints	0.75	682.7	LF	682.7	LF	512		

Table 7 MH_β = Estimate for Decreased Inlet Filter House Quantities

| Estimate—Inlet Filter House | | Historical | | | Estimate | | |
Description	MH	Quantity	Unit	Quantity	Unit	BM	IW	MW
Air Filter System								
Inlet filter house						**1557**	**0**	**0**
Sloped roof	8.40	1.9	Ton	1.9	Ton	16		
Hopper	20.00	0.9	Ton	0.9	Ton	18		
Pipe/tubing	1.94	44.7	LF	44.7	LF	87		
Hand rails	0.25	19.0	LF	19.0	LF	5		
Filter cartridges	0.75	918.0	EA	918.0	EA	689		
Filter modules (center)	8.40	14.0	Ton	14.0	Ton	117		
Filter modules (left and right hand)	8.40	8.3	Ton	8.3	Ton	70		
Walkway assembly (center)	8.40	4.2	Ton	4.2	Ton	35		
Walkway assembly (LH and RH)	8.40	3.1	Ton	3.1	Ton	26		
D15 weather hoods	8.40	8.9	Ton	8.9	Ton	75		
Field joints	0.75	558.5	LF	558.5	LF	419		

MHc is based on historical data from field erection of an F Class CTG in a combined cycle power plant and the scope and quantity differences can be identified and the impacts estimated therefore:

$MH_\beta <= MHc <= MH_\ni$

The proposed unit is based on the estimators quantity take off and erection man hours are either (+or−) 10%.

Direct proportion (straight line graph); comparison method quantities are directly proportional to estimate MHs.

MHc is directly proportional to MH_β and $MH_\ni$.

Whenever one variable increases or decreases, the other increases or decreases and vice versa.

Equation of straight line:

y is directly proportional to quantity x if y = kx for some k > 0.

Therefore, quantities are directly proportional to estimate MHs.

Unit Quantity Method

The method starts with the quantity take-off arranged in the erection sequence required to assemble and install the equipment. The estimator selects the task description by defining the work scope for the item to be installed. Each task is related to and performed by direct craft and divided into one or more subsystems that are further divided into assemblies made up of construction line items.

The unit-quantity model is given by the following:

$$MH_t = \sum n_i \left(MH_i \right)$$

where MH_t = man hours for equipment and piping field installation; n_i = task take-off quantity i; in dimensional units; MH_i = unit equipment/piping man hour associated with n_i; i = task 1,2,..., m from quantity takeoff associated w/field equipment/piping.

The n_i quantity is the takeoff for the field scope of work.

MH_i unit man hours are determined from piping and major equipment man hour tables.

The unit man hours were determined from historical data and they correspond to the labor productivity necessary to install the piping and equipment in an industrial construction facility.

The unit-quantity method allows a final cross-check of actual man hours to estimated man hours.

Factor:

$$MH_t = n_i \left(MH_i \right) \left(f_u \right)$$

where f_u = factor percent for productivity loss and alloy weld factor; u = 1,2,..., p. (Table 8).

Table 8 **Illustration of Unit Quantity Method**

Scope of Work—Quantity Takeoff	Quantity n_i		Man Hour MH_i	Total MH MH_t
Air Filter System				
Inlet filter house				
Sloped roof	2.1	Ton	8.40	18
Hopper	1.0	Ton	20.00	20
Pipe/tubing	49.7	LF	1.94	96
Hand rails	21.1	LF	0.25	5
Filter cartridges	1020.0	EA	0.75	765
Filter modules (center)	15.5	Ton	8.40	130
Filter modules (left and right hand)	9.3	Ton	8.40	78
Walkway Assembly (Center)	4.7	Ton	8.40	39
Walkway Assembly (LH and RH)	3.5	Ton	8.40	29
D15 weather hoods	9.9	Ton	8.40	83
Field joints	620.6	LF	0.75	465
Man Hour Total				**1730**

F Class CTG Installation Estimate.

2.3.3 Support Steel, Inlet Filter House, Transition Duct, Inlet Duct, Plenum Sheet

3. Estimate— inlet air filter.

Factors Affecting Labor Productivity and Impacts of Overtime

The American Association of Cost Engineers defines productivity as a "relative measure of labor efficiency, either good or bad, when compared to an established base or norm."

Estimates are based on a defined work scope, duration, start date, and clarifications and exceptions to the bid documents. The design may be incomplete or changes are made that will impact the bid estimate.

Examples of impacts that affect labor productivity:

Project has added work scope and owner request project completed on bid date. Request may require increased craft manpower, a second shift, overtime, and many other impacts to the schedule and estimate. The increased man hours, extended schedule, and other resources will impact the schedule and project cost.

Delay of owner-provided material and equipment to be installed will affect the erection sequence, duration, and schedule of work packages. Reassignment of craft is required.

These impacts will cause an increase in manpower and joint occupancy with other trades causing a drop in productivity.

Labor Factors

The estimator can use the following table of labor factors and values for factoring labor productivity for work packages.

The overtime labor factors provide a basis for estimating inefficiencies utilizing quantitative data actually measured on construction projects (Tables 9–11).

Table 9 Project Overtime Labor Factors

Week of Extended OT	50 h/week	60 h/week	72 h/week	84 h/week
1	0.05	0.09	0.14	0.25
2	0.07	0.12	0.2	0.30
3	0.08	0.14	0.27	0.35
4	0.09	0.19	0.32	0.40
5	0.15	0.24	0.37	0.45
6	0.16	0.28	0.42	0.50
7	0.24	0.33	0.46	0.53
8	0.23	0.36	0.49	0.56
9	0.26	0.38	0.50	0.57
10	0.28	0.39	0.51	0.58
11	0.28	0.40	0.52	0.59
12	0.29	0.41	0.53	0.60
13	0.31	0.44	0.54	0.61
14	0.32	0.45	0.55	0.62
15	0.33	0.46	0.56	0.63
16	0.34	0.47	0.57	0.64

Table 10 **Illustration for Project Labor Factors for Productivity Loss**

Total Craft Man hours Crew Size Man Hour Factor	Crew Size 25			Factor 1.00%	17,180 171.8 17,351.8 = Percent of Base MH 101.00%
Sub Total Overtime Factors		Crew Size	Impact	Factor	Additional Hours
Week	Hours of Exposure	MH>40	MH		
1	50	10	500	0.05	25
2	50	10	500	0.07	35
3	50	10	500	0.08	40
4	50	10	500	0.09	45
5	50	10	500	0.15	75
6	50	10	500	0.16	80
7					
8					
9					
10					
Sub total hours of exposure	300				
Sub Total—Additional Overtime Man Hours					300

Table 11 **Straight Time and Overtime Additional MHs**

Other Factors:		MH	Factor		1718
Straight Time	**MHs**	**17,180**	**10%**		**30**
Additional MHs					**19,400**
Overtime	**MHs**	**600**	**5%**		**2220 = Percent**
Additional MHs					**of MH**
					112.92%
	Range of Impact			**Use**	
Other Factors	**Minor**	**Average**	**Severe**	**ST**	**OT**
Stacking of trades	10%	20%	30%	10%	
Morale and attitude	5%	15%	30%		
Reassignment of manpower	5%	10%	15%		
Stacking of contractors own forces	10%	20%	30%		
Dilution of supervision	10%	15%	25%		
Learning curve	5%	15%	30%		
Errors and omissions	3%	5%	6%		
Sharing occupancy with owners operations	5%	15%	25%		
Joint occupancy with other trades	5%	12%	20%		5%
Interferences with access to work site	5%	12%	30%		
Deficient materials delivery	10%	25%	35%		
Fatigue	8%	10%	12%		
Ripple	8%	20%	30%		
Season/weather	10%	20%	30%		
Availability of skilled labor					
Ratio of name calls to open calls					
Plant permit requirements					
Safety					
				10%	5%

The principles and techniques in this manual will enable the experienced estimator to apply the methods to measure, analyze, and compile accurate estimates for industrial projects.

Equipment Section General Notes

Unload and Handling—Equipment includes unloading at erection site, cribbing, and moving to erection site.

Plates—Shim and set align and grout slide/base plates, set and grout fixators and makeup anchor bolts.

Lift, Set, Align—Assembly, setting, aligning and makeup of field joints.

Structural Steel—Man hours include handling, unloading, shaking out, erecting, and weld/bolt up connections.

Alignment—Equipment includes rough and final alignment.

Pumps—Man hours include handling, hauling, rigging, setting, and alignment of pump and motor.

Skids—Man hours include handling, hauling, rigging, picking, setting, and alignment.

Stack—Man hours include handling, hauling, rigging, setting, aligning, bolting/welding stack sections.

Electrical—Equipment does not include electrical man hours.

Insulation and Lagging—Equipment does not include man hours for insulation and lagging.

Start-Up—Equipment does not include start-up man hours.

Equipment Installation—Erection is for all components described in the scope of work to be erected in the field.

Process Piping

1

1.1 Section Introduction—Piping Schedules and Tables

This section contains schedules and tables that cover the complete labor for the field fabrication and installation of process piping in an industrial facility. The direct craft man hours for handling, welding, and bolting pipe is based on welding methods, pipe wall thickness, pressure, and temperature. The piping schedules for large bore pipe units for handling pipe are in diameter inch feet and the units for welding, bolting, and handling valves are in diameter inches. Piping man hour tables, for carbon steel piping, are derived from the schedules. The actual man hour production units were developed from job cost reports for process piping installed in the field on construction projects. The projects involved large crew sizes, a low ratio of regular employees and a safe work place that strives for zero accidents. Each of these factors has a negative effect on productivity and has been considered in the development of the piping man hours. Standardized man hours are required so that all work is from the same "base line" data. The base line can be modified and the standard man hours can be set up in a computerized estimating system. Man hours are for direct craft labor and do not include craft supervision and indirect craft support man hours. Refer to Schedule G—Alloy and nonferrous weld factors to apply the percentages for field fabrication and erection of alloy and nonferrous piping.

Industrial Piping and Equipment Estimating Manual. http://dx.doi.org/10.1016/B978-0-12-813946-2.00001-0

1.2   Piping Section General Notes

Handling and erecting pipe—man hours to unload, store in lay down, haul, rig, and align in place.
Field handling valves—man hours to place screwed, flanged and weld end valves, and expansion joints
Field handling control valves and specialty items—man hours are two times manual valves per valve
Field erection bolt-ups—man hours per joint to bolt-up valves, expansion joints, flanged fittings, and spools
Making on screwed fittings and valves—man hours for cutting, threading, handling, and erection per connection.
General Welding Notes
Manual butt welds—wall thickness of the pipe determines man hours that will apply per joint

> PWHT butt welds greater than or equal to 0.750″
>
> PWHT craft support—man hours for craft to warp/and remove pipe insulation for stress relief
>
> Apply percentages for welding alloys and nonferrous butt welds
>
> Preheating and stress-relieving butt welds are not included

Olet welds—man hours are two times butt weld and include cutting, placing, and welding per connection
Stub in weld—man hours are 1.5 times butt weld and include cutting, placing, and welding per connection
Socket welds—man hours include fit up and welding per joint
Hydro test pipe—man hours to place/remove blinds, open/close valves, removal/replacement of valves and specialty items and pipe sections as required, and drain lines after testing.

1.3   Schedule A—Combined Cycle Power Plant Piping

Standard Labor Estimating Units		
Facility—Combined Cycle Power Plant	**Large Bore Piping**	**Small Bore Piping**
	Unit of Measure	**Unit of Measure**
Description	**Man Hours per Unit**	**Man Hours per Unit**
Handle and Install Pipe, Carbon Steel, Welded Joint	**Diameter Inch Feet**	**MH/LF**
WT<=0.375″	0.07	0.18
0.406″<=WT<=0.500″	0.09	0.23
0.562″<=WT<=0.688″	0.11	0.28
0.718″<=WT<=0.938″	0.14	0.35
1.031″<=WT<=1.219″	0.20	0.50
1.250″<=WT<=1.312″	0.25	0.75
Welding Butt Welds, Carbon Steel, Arc-Uphill	**Diameter Inch**	**MH/EA**
WT<=0.375″	0.50	1.10
0.406″>=WT<=0.500″	0.55	1.20
0.562″>=WT<=0.688″	1.05	2.20
0.718″>=WT<=0.938″ (PWHT)	1.20	2.45
1.031″>=WT<=1.219″ (PWHT)	1.45	2.70
1.250″<=WT<=1.312″ (PWHT)	2.20	4.40
Olet-(SOL, TOL, WOL)	2×BW	2×BW
Stub in	1.5×BW	1.5×BW
Socketweld		Per SW table
PWHT craft support labor	0.45	1.00
Bolt-up of Flanged Joints by Weight Class	**Diameter Inch**	**MH/EA**
150#/300# bolt-up	0.40	1.00
600#/900# bolt-up	0.50	1.20
1500#/2500# bolt-up	0.65	1.60
Handle Valves by Weight Class	**Diameter Inch**	**MH/EA**
150# and 300# manual valve	0.45	1.00
600# and 900# manual valve	0.90	1.80
Heavier manual valve	1.00	2 x manual
Control Valve and Specialty Item	2.00	2 x manual

1.3.1 Handle and Install Pipe, Carbon Steel, Welded Joint

Facility—Combined Cycle Power Plant

Man Hours per Foot

	Wall Thickness Inches				
Pipe Size	0.375″ or Less	0.406″ to 0.500″	0.562″ to 0.688″	0.718″ to 0.938″	1.031″ to 1.219″
0.5	0.18	0.23	0.28	0.35	0.50
0.75	0.18	0.23	0.28	0.35	0.50
1	0.18	0.23	0.28	0.35	0.50
1.5	0.18	0.23	0.28	0.35	0.50
2	0.18	0.23	0.28	0.35	0.50
2.5	0.18	0.23	0.28	0.35	0.50
3	0.21	0.27	0.33	0.42	0.60
4	0.28	0.36	0.44	0.56	0.80
6	0.42	0.54	0.66	0.84	1.20
8	0.56	0.72	0.88	1.12	1.60
10	0.70	0.90	1.10	1.40	2.00
12	0.84	1.08	1.32	1.68	2.40
14	0.98	1.26	1.54	1.96	2.80
16	1.12	1.44	1.76	2.24	3.20
18	1.26	1.62	1.98	2.52	3.60
20	1.40	1.80	2.20	2.80	4.00
24	1.68	2.16	2.64	3.36	4.80

Erect pipe man hours include all labor to unload, store in lay down, haul, rig, and align in place.

1.3.2   *Welding Butt Weld, Carbon Steel, SMAW—Uphill*

Facility—Combined Cycle Power Plant

Man Hour per Joint

Pipe Size	Wall Thickness Inches				
	0.375″ or Less	0.406″ to 0.500″	0.562″ to 0.688″	0.718″ to 0.938″	1.031″ to 1.219″
0.5	1.10	1.20	2.20	2.45	2.70
0.75	1.10	1.20	2.20	2.45	2.70
1	1.10	1.20	2.20	2.45	2.70
1.5	1.10	1.20	2.20	2.45	2.70
2	1.10	1.20	2.20	2.45	2.70
2.5	1.25	1.38	2.63	3.00	3.63
3	1.50	1.65	3.15	3.60	4.35
4	2.00	2.20	4.20	4.80	5.80
6	3.00	3.30	6.30	7.20	**8.70**
8	4.00	4.40	8.40	**9.60**	**11.60**
10	5.00	5.50	10.50	**12.00**	**14.50**
12	6.00	6.60	12.60	**14.40**	**17.40**
14	7.00	7.70	14.70	**16.80**	**20.30**
16	8.00	8.80	16.80	**19.20**	**23.20**
18	9.00	9.90	18.90	**21.60**	**26.10**
20	10.00	11.00	21.00	**24.00**	**29.00**
24	12.00	13.20	25.20	**28.80**	**34.80**

Manual butt welds.
Wall thickness of the pipe determines man hours that will apply per joint.
PWHT butt welds greater than or equal to 0.750″.
Apply percentages for welding alloys and nonferrous butt welds.

1.3.3   Welding—SOL, TOL, WOL

Facility—Combined Cycle Power Plant

Man Hour per SOL, TOL, WOL

Pipe Size	Wall Thickness in Inches				
	0.375" or Less	0.406" to 0.500"	0.562" to 0.688"	0.718" to 0.938"	1.031" to 1.219"
0.5	2.20	2.40	4.40	4.90	5.40
0.75	2.20	2.40	4.40	4.90	5.40
1	2.20	2.40	4.40	4.90	5.40
1.5	2.20	2.40	4.40	4.90	5.40
2	2.20	2.40	4.40	4.90	5.40
2.5	2.50	2.75	5.25	6.00	7.25
3	3.00	3.30	6.30	7.20	8.70
4	4.00	4.40	8.40	9.60	11.60
6	6.00	6.60	12.60	14.40	**17.40**
8	8.00	8.80	16.80	**19.20**	**23.20**
10	10.00	11.00	21.00	**24.00**	**29.00**
12	12.00	13.20	25.20	**28.80**	**34.80**
14	14.00	15.40	29.40	**33.60**	**40.60**
16	16.00	17.60	33.60	**38.40**	**46.40**
18	18.00	19.80	37.80	**43.20**	**52.20**
20	20.00	22.00	42.00	**48.00**	**58.00**
24	24.00	26.40	50.40	**57.60**	**69.60**

SOL, TOL, WOL welds; two times BW (includes cut, prep, and fit up).
Wall thickness of the pipe used for olet determines man hours that will apply.
Apply percentages for welding alloys and nonferrous welds.

1.3.4 Welding Stub In

Facility—Combined Cycle Power Plant

Man Hour per Stub In

Pipe Size	Wall Thickness Inches				
	0.375″ or Less	0.406″ to 0.500″	0.562″ to 0.688″	0.718″ to 0.938″	1.031″ to 1.219″
0.5	1.65	1.80	3.30	3.68	4.05
0.75	1.65	1.80	3.30	3.68	4.05
1	1.65	1.80	3.30	3.68	4.05
1.5	1.65	1.80	3.30	3.68	4.05
2	1.65	1.80	3.30	3.68	4.05
2.5	1.88	2.06	3.94	4.50	5.44
3	2.25	2.48	4.73	5.40	6.53
4	3.00	3.30	6.30	7.20	8.70
6	4.50	4.95	9.45	10.80	**13.05**
8	6.00	6.60	12.60	**14.40**	**17.40**
10	7.50	8.25	15.75	**18.00**	**21.75**
12	9.00	9.90	18.90	**21.60**	**26.10**
14	10.50	11.55	22.05	**25.20**	**30.45**
16	12.00	13.20	25.20	**28.80**	**34.80**
18	13.50	14.85	28.35	**32.40**	**39.15**
20	15.00	16.50	31.50	**36.00**	**43.50**
24	18.00	19.80	37.80	**43.20**	**52.20**

Stub In 1.5 times BW (includes cut, prep, and fit up).
Wall thickness of the pipe used for stub in determines man hours that will apply.
Apply percentages for welding alloys and nonferrous welds.

1.3.5   PWHT Craft Support Labor

Facility—Combined Cycle Power Plant

Man Hour per Joint

Pipe Size	MH
0.5	1.00
0.75	1.00
1	1.00
1.5	1.00
2	1.00
2.5	1.13
3	1.35
4	1.80
6	2.70
8	3.60
10	4.50
12	5.40
14	6.30
16	7.20
18	8.10
20	9.00
24	10.80

Place and remove insulation from weld joint.

1.3.6   Field Bolt-up of Flanged Joints by Weight Class

Facility—Combined Cycle Power Plant

Man Hour per Joint

Pipe Size	Pressure Rating		
	150#/300#	600#/900#	1500#/2500#
0.5	1.00	1.20	1.60
0.75	1.00	1.20	1.60
1	1.00	1.20	1.60
1.5	1.00	1.20	1.60
2	1.00	1.20	1.60
2.5	1.00	1.25	1.63
3	1.20	1.50	1.95
4	1.60	2.00	2.60
6	2.40	3.00	3.90
8	3.20	4.00	5.20
10	4.00	5.00	6.50
12	4.80	6.00	7.80
14	5.60	7.00	9.10
16	6.40	8.00	10.40
18	7.20	9.00	11.70
20	8.00	10.00	13.00
24	9.60	12.00	15.60

Man hours per joint to bolt-up valves, expansion joints, flanged fittings, and spools.

1.3.7   Field Handle Valves/Specialty Items by Weight Class

Facility—Combined Cycle Power Plant

	Man Hours per Valve/Specialty			Control Valve/Specialty		
	Pressure Rating			Pressure Rating		
Pipe Size	150#/300#	600#/900#	1500#/2500#	150#/300#	600#/900#	1500#/2500#
0.5	1.00	1.80	2.00	2.00	3.60	4.00
0.8	1.00	1.80	2.00	2.00	3.60	4.00
1	1.00	1.80	2.00	2.00	3.60	4.00
1.5	1.00	1.80	2.00	2.00	3.60	4.00
2	1.00	1.80	2.00	2.00	3.60	4.00
2.5	1.13	2.25	4.50	2.25	4.50	9.00
3	1.35	2.70	5.40	2.70	5.40	10.80
4	1.80	3.60	7.20	3.60	7.20	14.40
6	2.70	5.40	10.80	5.40	10.80	21.60
8	3.60	7.20	14.40	7.20	14.40	28.80
10	4.50	9.00	18.00	9.00	18.00	36.00
12	5.40	10.80	21.60	10.80	21.60	43.20
14	6.30	12.60	25.20	12.60	25.20	50.40
16	7.20	14.40	28.80	14.40	28.80	57.60
18	8.10	16.20	32.40	16.20	32.40	64.80
20	9.00	18.00	36.00	18.00	36.00	72.00
24	10.80	21.60	43.20	21.60	43.20	86.40

Field handle and erect screwed, flanged, weld end valves, and expansion joints.

1.4 Schedule B—Simple Cycle Power Plant Piping

Standard Labor Estimating Units		
Facility—Simple Cycle Power Plant	**Large Bore Piping**	**Small Bore Piping**
	Unit of Measure	**Unit of Measure**
Description	**Man Hours per Unit**	**Man Hours per Unit**
	Diameter Inch Feet	**MH/LF**
Handle and Install S10 and Lighter S.S. Pipe	0.06	0.15
Handle and Install Pipe, Carbon Steel	**Diameter Inch Feet**	**MH/LF**
WT <= 0.375″	0.06	0.15
0.406″ >= WT <= 0.500″	0.07	0.18
Welding, Carbon Steel, Butt weld (Arc-Uphill)	**Diameter Inch**	**MH/Joint**
WT <= 0.375″	0.50	1.00
0.406″ >= WT <= 0.500″	0.60	1.10
SOL, TOL, WOL	2 × BW	2 × BW
Stub in	1.5 × BW	1.5 × BW
Socketweld		Per SW table
Welding, Stainless Steel, Butt weld (Heliarc)	**Diameter Inch**	**MH/Joint**
Schedule 10S and lighter	0.70	1.54
Bolt-up of Flanged Joints by Weight Class	**Diameter Inch**	**MH/EA**
150#/300# bolt-up	0.40	1.00
600#/900# bolt-up	0.60	1.20
Handle Valves by Weight Class	**Diameter Inch**	**MH/EA**
150# and 300# manual valve	0.45	1.00
600# and 900# manual valve	0.65	1.30

1.4.1   S10S and Lighter Stainless Steel Piping

Facility—Simple Cycle Power Plant

Pipe Size	Handle Pipe MH/LF	BW (Heliarc) MH/Joint	Man Hour per Bolt-up		Man Hour per Valve	
			Pressure Rating		Pressure Rating	
			150#/300#	600#/900#	150#/300#	600#/900#
0.5	0.15	1.54	1.00	1.20	1.00	1.30
0.8	0.15	1.54	1.00	1.20	1.00	1.30
1	0.15	1.54	1.00	1.20	1.00	1.30
1.5	0.15	1.54	1.00	1.20	1.00	1.30
2	0.15	1.54	1.00	1.20	1.00	1.30
2.5	0.15	1.75	1.00	1.50	1.13	1.63
3	0.18	2.10	1.20	1.80	1.35	1.95
4	0.24	2.80	1.60	2.40	1.80	2.60
6	0.36	4.20	2.40	3.60	2.70	3.90
8	0.48	5.60	3.20	4.80	3.60	5.20
10	0.60	7.00	4.00	6.00	4.50	6.50
12	0.72	8.40	4.80	7.20	5.40	7.80
14	0.84	9.80	5.60	8.40	6.30	9.10
16	0.96	11.20	6.40	9.60	7.20	10.40
18	1.08	12.60	7.20	10.80	8.10	11.70
20	1.20	14.00	8.00	12.00	9.00	13.00
24	1.44	16.80	9.60	14.40	10.80	15.60

Man hours for pipe handling include all labor to unload, store in lay down, haul, and align in place.
Man hours per joint to bolt-up valves, expansion joints, flanged fittings, and spools.
Field handle and erect screwed, flanged, weld end valves, and expansion joints.
Manual butt welds.
Wall thickness of the pipe determines man hours that will apply per joint.

1.4.2   Carbon Steel Pipe, Welded Joint

Facility—Simple Cycle Power Plant

Man Hour per Foot

	Handle Pipe		Man Hours Weld Joint		Man Hours per SOL, TOL, WOL	
	Wall Thickness in Inches					
Pipe Size	0.375″ or Less	0.406″ to 0.500″	0.375″ or Less	0.406″ to 0.500″	0.375″ or Less	0.406″ to 0.500″
0.5	0.15	0.18	1.00	1.10	2.00	2.20
0.8	0.15	0.18	1.00	1.10	2.00	2.20
1	0.15	0.18	1.00	1.10	2.00	2.20
1.5	0.15	0.18	1.00	1.10	2.00	2.20
2	0.15	0.18	1.00	1.10	2.00	2.20
2.5	0.15	0.18	1.25	1.50	2.50	3.00
3	0.18	0.21	1.50	1.80	3.00	3.60
4	0.24	0.28	2.00	2.40	4.00	4.80
6	0.36	0.42	3.00	3.60	6.00	7.20
8	0.48	0.56	4.00	4.80	8.00	9.60
10	0.60	0.70	5.00	6.00	10.00	12.00
12	0.72	0.84	6.00	7.20	12.00	14.40
14	0.84	0.98	7.00	8.40	14.00	16.80
16	0.96	1.12	8.00	9.60	16.00	19.20
18	1.08	1.26	9.00	10.80	18.00	21.60
20	1.20	1.40	10.00	12.00	20.00	24.00
24	1.44	1.68	12.00	14.40	24.00	28.80

Man hours for pipe handling include all labor to unload, store in lay down, haul, and align in place.
Manual butt welds.
Wall thickness of the pipe determines man hours that will apply per joint.
Apply percentages for welding alloys and nonferrous butt welds.
Preheating and stress-relieving butt welds are not included.

1.4.3 Field Bolt-up of Flanged Joints by Weight Class

Facility—Simple Cycle Power Plant

Man Hours per Joint

Pipe Size	Pressure Rating	
	150#/300#	600#/900#
0.5	1.00	1.30
0.8	1.00	1.30
1	1.00	1.30
1.5	1.00	1.30
2	1.00	1.30
2.5	1.00	1.50
3	1.20	1.80
4	1.60	2.40
6	2.40	3.60
8	3.20	4.80
10	4.00	6.00
12	4.80	7.20
14	5.60	8.40
16	6.40	9.60
18	7.20	10.80
20	8.00	12.00
24	9.60	14.40

Man hours per joint to bolt-up valves, expansion joints, flanged fittings, and spools.

1.4.4   Field Handle Valves/Specialty Items by Weight Class

Facility—Simple Cycle Power Plant

Pipe Size	Man Hours per Valve/Specialty		Control Valve/Specialty	
	Pressure Rating		Pressure Rating	
	150#/300#	600#/900#	150#/300#	600#/900#
0.5	1.00	1.30	2.00	2.6
0.75	1.00	1.30	2.00	2.6
1	1.00	1.30	2.00	2.6
1.5	1.00	1.30	2.00	2.6
2	1.00	1.30	2.00	2.6
2.5	1.13	1.63	2.25	3.25
3	1.35	1.95	2.70	3.9
4	1.80	2.60	3.60	5.2
6	2.70	3.90	5.40	7.8
8	3.60	5.20	7.20	10.4
10	4.50	6.50	9.00	13
12	5.40	7.80	10.80	15.6
14	6.30	9.10	12.60	18.2
16	7.20	10.40	14.40	20.8
18	8.10	11.70	16.20	23.4
20	9.00	13.00	18.00	26
24	10.80	15.60	21.60	31.2

Field handle and erect screwed, flanged, weld end valves, and expansion joint.

1.5   Schedule C—Refinery Piping, Hydrogen Plant

Facility—Oil Refinery, Hydrogen Plant

Standard Labor Estimating Units		
Facility—Oil Refinery, Hydrogen Plant	**Large Bore Piping**	**Small Bore Piping**
	Unit of Measure	**Unit of Measure**
Description	**Man Hours per Unit**	**Man Hours per Unit**
Handle and Install Pipe, Carbon Steel, Welded Joint	**Diameter Inch Feet**	**MH/LF**
WT<=0.375″	0.06	0.15
0.406″>=WT<=0.500″	0.07	0.18
0.562″>=WT<=0.688″	0.08	0.20
0.718″<=WT<=0.938″	0.11	0.28
1.031″<=WT<=1.219″	0.20	0.50
1.250″<=WT<=1.312″	0.25	0.75
Welding, Carbon Steel, Arc-Uphill	**Diameter Inch**	**MH/EA**
WT<=0.375″	0.45	1.00
0.406″>=WT<=0.500″	0.50	1.15
0.562″>=WT<=0.688″	0.95	1.90
0.718″>=WT<=0.938″ (PWHT)	1.10	2.25
1.031″>=WT<=1.218″ (PWHT)	1.35	2.60
1.250″<=WT<=1.312″ (PWHT)	2.10	4.20
SOL, TOL, WOL	2×BW	2×BW
Stub in	1.5×BW	1.5×BW
Socketweld		Per SW table
PWHT craft support labor	0.45	1.00
Bolt-up of Flanged Joints by Weight Class	**Diameter Inch**	**MH/EA**
150#/300# bolt-up	0.40	1.00
600#/900# bolt-up	0.50	1.20
1500# bolt-up	0.65	1.60
Handle Valves by Weight Class	**Diameter Inch**	**MH/EA**
150# and 300# manual valve	0.50	1.00
600# and 900# manual valve	0.52	1.30
Heavier manual valve>=1500#	1.00	2.00
Control valve and specialty items	2 X manual	2 X manual

1.5.1   Handle and Install Pipe, Carbon Steel, Welded Joint

Facility—Oil Refinery, Hydrogen Plant

Man Hours per Foot

Pipe Size	Wall Thickness in Inches				
	Up to 0.375″	0.406″ to 0.500″	0.562″ to 0.688″	0.718″ to 0.938″	1.031″ to 1.219″
0.5	0.15	0.18	0.20	0.28	0.50
0.75	0.15	0.18	0.20	0.28	0.50
1	0.15	0.18	0.20	0.28	0.50
1.5	0.15	0.18	0.20	0.28	0.50
2	0.15	0.18	0.20	0.28	0.50
2.5	0.15	0.18	0.20	0.28	0.50
3	0.18	0.21	0.24	0.33	0.60
4	0.24	0.28	0.32	0.44	0.80
6	0.36	0.42	0.48	0.66	1.20
8	0.48	0.56	0.64	0.88	1.60
10	0.60	0.70	0.80	1.10	2.00
12	0.72	0.84	0.96	1.32	2.40
14	0.84	0.98	1.12	1.54	2.80
16	0.96	1.12	1.28	1.76	3.20
18	1.08	1.26	1.44	1.98	3.60
20	1.20	1.40	1.60	2.20	4.00
24	1.44	1.68	1.92	2.64	4.80

Erect pipe man hours include all labor to unload, store in lay down, haul, rig, and align in place.

1.5.2   *Welding Butt Weld, Carbon Steel, SMAW—Uphill*

Facility—Oil Refinery, Hydrogen Plant

Man Hours per Joint

Pipe Size	Wall Thickness in Inches				
	Up to 0.375″	0.406″ to 0.500″	0.562″ to 0.688″	0.718″ to 0.938″	1.031″ to 1.219″
0.5	1.00	1.15	1.90	2.25	2.60
0.75	1.00	1.15	1.90	2.25	2.60
1	1.00	1.15	1.90	2.25	2.60
1.5	1.00	1.15	1.90	2.25	2.60
2	1.00	1.15	1.90	2.25	2.60
2.5	1.13	1.25	2.38	2.75	3.38
3	1.35	1.50	2.85	3.30	4.05
4	1.80	2.00	3.80	4.40	5.40
6	2.70	3.00	5.70	6.60	**8.10**
8	3.60	4.00	7.60	**8.80**	**10.80**
10	4.50	5.00	9.50	**11.00**	**13.50**
12	5.40	6.00	11.40	**13.20**	**16.20**
14	6.30	7.00	13.30	**15.40**	**18.90**
16	7.20	8.00	15.20	**17.60**	**21.60**
18	8.10	9.00	17.10	**19.80**	**24.30**
20	9.00	10.00	19.00	**22.00**	**27.00**
24	10.80	12.00	22.80	**26.40**	**32.40**

Manual butt welds.
Wall thickness of the pipe determines man hours that will apply per joint.
PWHT butt welds greater than or equal to 0.750″.
Apply percentages for welding alloys and nonferrous butt welds.

1.5.3 Welding-SOL, TOL, WOL

Facility—Oil Refinery, Hydrogen Plant

Man Hours per SOL, TOL, WOL

Pipe Size	Wall Thickness in Inches				
	Up to 0.375″	0.406″ to 0.500″	0.562″ to 0.688″	0.718″ to 0.938″	1.031″ to 1.219″
0.5	2.00	2.30	3.80	4.50	5.20
0.75	2.00	2.30	3.80	4.50	5.20
1	2.00	2.30	3.80	4.50	5.20
1.5	2.00	2.30	3.80	4.50	5.20
2	2.00	2.30	3.80	4.50	5.20
2.5	2.25	2.50	4.75	5.50	6.75
3	2.70	3.00	5.70	6.60	8.10
4	3.60	4.00	7.60	8.80	10.80
6	5.40	6.00	11.40	13.20	**16.20**
8	7.20	8.00	15.20	**17.60**	**21.60**
10	9.00	10.00	19.00	**22.00**	**27.00**
12	10.80	12.00	22.80	**26.40**	**32.40**
14	12.60	14.00	26.60	**30.80**	**37.80**
16	14.40	16.00	30.40	**35.20**	**43.20**
18	16.20	18.00	34.20	**39.60**	**48.60**
20	18.00	20.00	38.00	**44.00**	**54.00**
24	21.60	24.00	45.60	**52.80**	**64.80**

SOL, TOL, WOL welds; two times BW (includes cut, prep, and fit up).
Wall thickness of the pipe used for weld determines man hours that will apply.
Apply percentages for welding alloys and nonferrous welds.

1.5.4   PWHT Craft Support Labor

Facility—Oil Refinery, Hydrogen Plant

Man Hour per Joint

Pipe Size	MH
0.5	1.00
0.75	1.00
1	1.00
1.5	1.00
2	1.00
2.5	1.13
3	1.35
4	1.80
6	2.70
8	3.60
10	4.50
12	5.40
14	6.30
16	7.20
18	8.10
20	9.00
24	10.80

Place and remove insulation from weld joint.

1.5.5   Field Bolt-up of Flanged Joints by Weight Class

Facility—Oil Refinery, Hydrogen Plant

Man Hours per Joint

Pipe Size	Pressure Rating		
	150#/300#	600#/900#	1500#/2500#
0.5	1.00	1.20	1.60
0.75	1.00	1.20	1.60
1	1.00	1.20	1.60
1.5	1.00	1.20	1.60
2	1.00	1.20	1.60
2.5	1.00	1.25	1.63
3	1.20	1.50	1.95
4	1.60	2.00	2.60
6	2.40	3.00	3.90
8	3.20	4.00	5.20
10	4.00	5.00	6.50
12	4.80	6.00	7.80
14	5.60	7.00	9.10
16	6.40	8.00	10.40
18	7.20	9.00	11.70
20	8.00	10.00	13.00
24	9.60	12.00	15.60

Man hours per joint to bolt-up valves, expansion joints, flanged fittings, and spools.

1.5.6   Field Handle Valves/Specialty Items by Weight Class

Facility—Oil Refinery, Hydrogen Plant

Pipe Size	Man Hours per Valve/Specialty			Control Valve/Specialty		
	Manual Valve Pressure Rating			Pressure Rating		
	150#/300#	600#/900#	1500#/2500#	150#/300#	600#/900#	1500#/2500#
0.5	1.00	1.30	2.00	2.00	2.60	4.00
0.75	1.00	1.30	2.00	2.00	2.60	4.00
1	1.00	1.30	2.00	2.00	2.60	4.00
1.5	1.00	1.30	2.00	2.00	2.60	4.00
2	1.00	1.30	2.00	2.00	2.60	4.00
2.5	1.25	1.30	2.50	2.50	2.60	5.00
3	1.50	1.56	3.00	3.00	3.12	6.00
4	2.00	2.08	4.00	4.00	4.16	8.00
6	3.00	3.12	6.00	6.00	6.24	12.00
8	4.00	4.16	8.00	8.00	8.32	16.00
10	5.00	5.20	10.00	10.00	10.40	20.00
12	6.00	6.24	12.00	12.00	12.48	24.00
14	7.00	7.28	14.00	14.00	14.56	28.00
16	8.00	8.32	16.00	16.00	16.64	32.00
18	9.00	9.36	18.00	18.00	18.72	36.00
20	10.00	10.40	20.00	20.00	20.80	40.00
24	12.00	12.48	24.00	24.00	24.96	48.00

Man hours only-field handle and erect screwed, flanged, weld end valves, and expansion joints.

1.6   Schedule D—Petroleum Tank Farm Piping

Standard Labor Estimating Units		
Facility—Petroleum Industry Tank Farm	Large Bore Piping	Small Bore Piping
	Unit of Measure	Unit of Measure
Description	Man Hours per Unit	Man Hours per Unit
Handle and Install Pipe, Carbon Steel, Welded Joint	Diameter Inch Feet	**MH/LF**
WT<=0.375″	0.04	0.12
0.406″>=WT<=0.500″	0.05	0.14
0.562″>=WT<=0.688″	0.06	0.15
Welding, Carbon Steel, Arc-Downhill	Diameter Inch	**MH/EA**
WT<=0.375″	0.35	0.80
0.406″>=WT<=0.500″	0.45	1.00
0.562″>=WT<=0.688″	0.80	1.60
SOL, TOL, WOL	2×BW	2×BW
Stub in	1.5×BW	
Socket welds	0.65	1.00
Bolt-up of Flanged Joints by Weight Class	Diameter Inch	**MH/EA**
150#/300# bolt-up	0.40	1.00
600# bolt-up	0.60	1.30
Handle Valves by Weight Class	Diameter Inch	**MH/Valve**
150# and 300# manual valve	0.35	0.80
600# manual valve	0.58	1.30
Control valve and specialty items	2 X manual	2 X manual

1.6.1 Handle and Install Pipe, Carbon Steel, Welded Joint

Facility—Petroleum Industry Tank Farm

Man Hours per Foot

Pipe Size	Wall Thickness in Inches		
	0.375″ or less	0.406″ to 0.500″	0.562″ to 0.688″
0.5	0.12	0.14	0.15
0.75	0.12	0.14	0.15
1	0.12	0.14	0.15
1.5	0.12	0.14	0.15
2	0.12	0.14	0.15
2.5	0.10	0.13	0.15
3	0.12	0.15	0.18
4	0.16	0.20	0.24
6	0.24	0.30	0.36
8	0.32	0.40	0.48
10	0.40	0.50	0.60
12	0.48	0.60	0.72
14	0.56	0.70	0.84
16	0.64	0.80	0.96
18	0.72	0.90	1.08
20	0.80	1.00	1.20
24	0.96	1.20	1.44

Erect pipe man hours include all labor to unload, store in lay down, haul, rig, and align in place.

1.6.2 Welding Butt Weld, Carbon Steel, SMAW—Downhill

Facility—Petroleum Industry Tank Farm

Man Hours Weld Joint

Pipe Size	Wall Thickness in Inches		
	0.375″ or less	0.406″ to 0.500″	0.562″ to 0.688″
0.5	0.80	1.00	1.60
0.75	0.80	1.00	1.60
1	0.80	1.00	1.60
1.5	0.80	1.00	1.60
2	0.80	1.00	1.60
2.5	0.88	1.13	2.00
3	1.05	1.35	2.40
4	1.40	1.80	3.20
6	2.10	2.70	4.80
8	2.80	3.60	6.40
10	3.50	4.50	8.00
12	4.20	5.40	9.60
14	4.90	6.30	11.20
16	5.60	7.20	12.80
18	6.30	8.10	14.40
20	7.00	9.00	16.00
24	8.40	10.80	19.20

Manual butt welds.
PWHT butt welds greater than or equal to 0.750″.
Apply percentages for welding alloys and nonferrous butt welds.
Preheating and stress-relieving butt welds are not included.

1.6.3   Welding—SOL, TOL, WOL

Facility—Petroleum Tank Farm Piping

Man Hours per—SOL, TOL, WOL

Pipe Size	Wall Thickness in Inches		
	0.375″ or less	0.406′ to 0.500′	0.562′ to 0.688′
0.5	1.60	2.00	3.20
0.75	1.60	2.00	3.20
1	1.60	2.00	3.20
1.5	1.60	2.00	3.20
2	1.60	2.00	3.20
2.5	1.75	2.25	4.00
3	2.10	2.70	4.80
4	2.80	3.60	6.40
6	4.20	5.40	9.60
8	5.60	7.20	12.80
10	7.00	9.00	16.00
12	8.40	10.80	19.20
14	9.80	12.60	22.40
16	11.20	14.40	25.60
18	12.60	16.20	28.80
20	14.00	18.00	32.00
24	16.80	21.60	38.40

SOL, TOL, WOL welds; two times BW (includes cut, prep, and fit up).
Wall thickness of the pipe used for weld determines man hours that will apply.
Apply percentages for welding alloys and nonferrous welds.
Preheating and stress-relieving welds are not included.

1.6.4   Field Bolt-up of Flanged Joints by Weight Class

Facility—Petroleum Industry Tank Farm

Man Hours per Joint

Pipe Size	Pressure Rating	
	150#/300#	600#
0.5	1.00	1.30
0.75	1.00	1.30
1	1.00	1.30
1.5	1.00	1.30
2	1.00	1.30
2.5	1.00	1.50
3	1.20	1.80
4	1.60	2.40
6	2.40	3.60
8	3.20	4.80
10	4.00	6.00
12	4.80	7.20
14	5.60	8.40
16	6.40	9.60
18	7.20	10.80
20	8.00	12.00
24	9.60	14.40

Man hours per joint to bolt-up valves, expansion joints, flanged fittings, and spools.

1.6.5 Field Handle Valves/Specialty Items by Weight Class

Facility—Petroleum Industry Tank Farm

Pipe Size	Man Hours per Valve/Specialty		Control Valve/Specialty	
	Valve Pressure Rating		Pressure Rating	
	150#/300#	600#	150#/300#	600#
0.5	0.80	1.30	1.60	2.60
0.75	0.80	1.30	1.60	2.60
1	0.80	1.30	1.60	2.60
1.5	0.80	1.30	1.60	2.60
2	0.80	1.30	1.60	2.60
2.5	0.88	1.45	1.75	2.90
3	1.05	1.74	2.10	3.48
4	1.40	2.32	2.80	4.64
6	2.10	3.48	4.20	6.96
8	2.80	4.64	5.60	9.28
10	3.50	5.80	7.00	11.60
12	4.20	6.96	8.40	13.92
14	4.90	8.12	9.80	16.24
16	5.60	9.28	11.20	18.56
18	6.30	10.44	12.60	20.88
20	7.00	11.60	14.00	23.20
24	8.40	13.92	16.80	27.84

Field handle and erect screwed, flanged, weld end valves, and expansion joint.

1.7   Schedule E—Brewery, Food Processing Plants

Standard Labor Estimating Units		
Facility—Brewery, Food Processing Plants	Large Bore Piping	Small Bore Piping
	Unit of Measure	Unit of Measure
Description	Man Hours per Unit	Man Hours per Unit
	Diameter Inch Feet	MH/LF
Handle and **Install S10** and Lighter S.S. Pipe	0.06	0.15
Handle and Install Pipe, Carbon Steel	Diameter Inch Feet	MH/LF
WT<=0.375″	0.04	0.12
0.406″>=WT<=0.500″	0.06	0.15
Welding, Carbon Steel, Butt Weld (Arc-Uphill)	Diameter Inch	MH/JT
WT<=0.375″	0.45	1.00
0.406″>=WT<=0.500″	0.60	1.10
SOL, TOL, WOL	2×BW	2×BW
Stub in	2×BW	2×BW
Socket welds	0.65	
Welding, Stainless Steel, Butt Weld (TIG)	Diameter Inch	MH/JT
Schedule 10S and lighter (not polished)	0.65	1.30
Bolt-up of Flanged Joints by Weight Class	Diameter Inch	MH/EA
150#/300# bolt-up	0.40	1.00
600# bolt-up	0.60	1.30
Handle Valves by Weight Class	Diameter Inch	MH/Valve
150# and 300# manual valve	0.32	0.80
600# manual valve	0.60	1.30

1.7.1 S10S and Lighter Stainless Steel Piping

Facility—Brewery, Food Processing Plant

Pipe Size	Handle Pipe MH/LF	BW (TIG) MH/Joint
0.5	0.15	1.30
0.75	0.15	1.30
1	0.15	1.30
1.5	0.15	1.30
2	0.15	1.30
2.5	0.15	1.63
3	0.18	1.95
4	0.24	2.60
6	0.36	3.90
8	0.48	5.20
10	0.60	6.50
12	0.72	7.80
14	0.84	9.10
16	0.96	10.40
18	1.08	11.70
20	1.20	13.00
24	1.44	15.60

Erect pipe man hours include all labor to unload, store in lay down, haul, rig, and align in place.
Manual butt welds.
Wall thickness of the pipe determines man hours that will apply per joint.
Apply percentages for welding alloys and nonferrous butt welds.
Preheating and stress-relieving butt welds are not included.

1.7.2 Carbon Steel Pipe, Welded Joint

Facility—Brewery, Food Processing Plant

	Man Hours per Foot		Man Hours Weld Joint		MH per SOL, TOL, WOL	
	Handle Pipe		Wall Thickness in Inches			
Pipe Size	0.375″ or Less	0.406′ to 0.500′	0.375″ or Less	0.406′ to 0.500′	0.375″ or Less	0.406′ to 0.500′
0.5	0.12	0.15	1.00	1.10	2.00	2.20
0.75	0.12	0.15	1.00	1.10	2.00	2.20
1	0.12	0.15	1.00	1.10	2.00	2.20
1.5	0.12	0.15	1.00	1.10	2.00	2.20
2	0.12	0.15	1.00	1.10	2.00	2.20
2.5	0.10	0.15	1.13	1.50	2.25	3.00
3	0.12	0.18	1.35	1.80	2.70	3.60
4	0.16	0.24	1.80	2.40	3.60	4.80
6	0.24	0.36	2.70	3.60	5.40	7.20
8	0.32	0.48	3.60	4.80	7.20	9.60
10	0.40	0.60	4.50	6.00	9.00	12.00
12	0.48	0.72	5.40	7.20	10.80	14.40
14	0.56	0.84	6.30	8.40	12.60	16.80
16	0.64	0.96	7.20	9.60	14.40	19.20
18	0.72	1.08	8.10	10.80	16.20	21.60
20	0.80	1.20	9.00	12.00	18.00	24.00
24	0.96	1.44	10.80	14.40	21.60	28.80

Erect pipe man hours include all labor to unload, store in lay down, haul, rig, and align in place.
Manual butt welds.
Wall thickness of the pipe determines man hours that will apply per joint.
Apply percentages for welding alloys and nonferrous butt welds.
Preheating and stress-relieving butt welds are not included.

1.7.3   Field Bolt-up of Flanged Joints by Weight Class

Facility—Brewery, Food Processing Plant

Pipe Size	Man Hours per Joint	
	Pressure Rating	
	150#/300#	600#
0.5	1.00	1.30
0.75	1.00	1.30
1	1.00	1.30
1.5	1.00	1.30
2	1.00	1.30
2.5	1.00	1.50
3	1.20	1.80
4	1.60	2.40
6	2.40	3.60
8	3.20	4.80
10	4.00	6.00
12	4.80	7.20
14	5.60	8.40
16	6.40	9.60
18	7.20	10.80
20	8.00	12.00
24	9.60	14.40

Man hours per joint to bolt-up valves, expansion joints, flanged fittings, and spools.

1.7.4   Field Handle Valves/Specialty Items by Weight Class

Facility—Brewery, Food Processing Plant

Pipe Size	Man Hours per Valve/Specialty		Control Valve/Specialty	
	Valve Pressure Rating		Pressure Rating	
	150#/300#	600#	150#/300#	600#
0.5	0.80	1.30	1.60	2.60
0.75	0.80	1.30	1.60	2.60
1	0.80	1.30	1.60	2.60
1.5	0.80	1.30	1.60	2.60
2	0.80	1.30	1.60	2.60
2.5	0.80	1.50	1.60	3.00
3	0.96	1.80	1.92	3.60
4	1.28	2.40	2.56	4.80
6	1.92	3.60	3.84	7.20
8	2.56	4.80	5.12	9.60
10	3.20	6.00	6.40	12.00
12	3.84	7.20	7.68	14.40
14	4.48	8.40	8.96	16.80
16	5.12	9.60	10.24	19.20
18	5.76	10.80	11.52	21.60
20	6.40	12.00	12.80	24.00
24	7.68	14.40	15.36	28.80

Field handle and erect screwed, flanged, weld end valves, and expansion joints.

1.8   Schedule F—Pump and Compressor Station Piping

Standard Labor Estimating Units		
Facility—Pump and Compressor Station Piping	Large Bore Piping	Small Bore Piping
	Unit of Measure	Unit of Measure
Description	Man Hours per Unit	Man Hours per Unit
Handle and Install Pipe, Carbon Steel, Welded Joint	Diameter Inch Feet	MH/LF
WT <= 0.375″	0.05	0.12
0.406″ >= WT <= 0.500″	0.06	0.15
0.562″ >= WT <= 0.688″	0.07	0.18
0.718″ <= WT <= 0.938″	0.10	0.25
1.031″ <= WT <= 1.219″	0.12	0.30
1.250″ <= WT <= 1.312″	0.15	0.35
Welding, Carbon Steel, Arc-Downhill	Diameter Inch Feet	MH/EA
WT <= 0.375″	0.35	0.80
0.406″ >= WT <= 0.500″	0.40	1.00
0.562″ >= WT <= 0.688″	0.85	1.90
0.718″ >= WT <= 0.938″ (PWHT)	1.15	2.40
1.031″ >= WT <= 1.218″ (PWHT)	1.40	2.90
1.250″ <= WT <= 1.312″ (PWHT)	1.50	3.00
SOL, TOL, WOL	2 × BW	
Stub in	1.5 × BW	
Socketweld	0.65	
PWHT craft support labor	0.45	1.00
Bolt-up of Flanged Joints by Weight Class	Diameter Inch Feet	MH/EA
150#/300# bolt-up	0.45	0.90
600#/900# bolt-up	0.55	1.20
1500# bolt-up	0.65	1.60
Handle Valves by Weight Class	Diameter Inch Feet	MH/EA
150# and 300# manual valve	0.35	0.80
600# and 900# manual valve	0.60	1.30
Heavier manual valve >= 1500#	0.90	1.80
Control valve and specialty items	2 X manual	

1.8.1 Handle and Install Pipe, Carbon Steel, Welded Joint

Facility—Pump and Compressor Station Piping

Man Hours per Foot

Pipe Size	Wall Thickness in Inches				
	0.375" or Less	0.406" to 0.500"	0.562" to 0.688"	0.718" to 0.938"	1.031" to 1.219"
0.5	0.12	0.15	0.18	0.25	0.30
0.75	0.12	0.15	0.18	0.25	0.30
1	0.12	0.15	0.18	0.25	0.30
1.5	0.12	0.15	0.18	0.25	0.30
2	0.12	0.15	0.18	0.25	0.30
2.5	0.13	0.15	0.18	0.25	0.30
3	0.15	0.18	0.21	0.30	0.36
4	0.20	0.24	0.28	0.40	0.48
6	0.30	0.36	0.42	0.60	0.72
8	0.40	0.48	0.56	0.80	0.96
10	0.50	0.60	0.70	1.00	1.20
12	0.60	0.72	0.84	1.20	1.44
14	0.70	0.84	0.98	1.40	1.68
16	0.80	0.96	1.12	1.60	1.92
18	0.90	1.08	1.26	1.80	2.16
20	1.00	1.20	1.40	2.00	2.40
24	1.20	1.44	1.68	2.40	2.88

Erect pipe man hours include all labor to unload, store in lay down, haul, rig, and align in place.

1.8.2   *Welding Butt Weld, Carbon Steel, SMAW—Downhill*

Facility—Pump and Compressor Station Piping

Man Hours Weld Joint

Pipe Size	Wall Thickness in Inches				
	0.375″ or Less	0.406″ to 0.500″	0.562″ to 0.688″	0.718″ to 0.938″	1.031″ to 1.219″
0.5	0.80	1.00	1.90	2.40	2.90
0.75	0.80	1.00	1.90	2.40	2.90
1	0.80	1.00	1.90	2.40	2.90
1.5	0.80	1.00	1.90	2.40	2.90
2	0.80	1.00	1.90	2.40	2.90
2.5	0.88	1.00	2.13	2.88	3.50
3	1.05	1.05	1.20	3.45	3.45
4	1.40	1.40	1.60	3.40	4.60
6	2.10	2.10	2.40	5.10	**6.90**
8	2.80	2.80	3.20	**6.80**	**9.20**
10	3.50	3.50	4.00	**8.50**	**11.50**
12	4.20	4.20	4.80	**10.20**	**13.80**
14	4.90	4.90	5.60	**11.90**	**16.10**
16	5.60	5.60	6.40	**13.60**	**18.40**
18	6.30	6.30	7.20	**15.30**	**20.70**
20	7.00	7.00	8.00	**17.00**	**23.00**
24	8.40	8.40	9.60	**20.40**	**27.60**

Manual butt welds.
Wall thickness of the pipe determines man hours that will apply per joint.
PWHT butt welds greater than or equal to 0.750″.
Apply percentages for welding alloys and nonferrous butt welds.
Preheating and stress-relieving butt welds are not included.

1.8.3   Welding—SOL, TOL, WOL

Facility—Pump and Compressor Station Piping

Man Hours per-SOL, TOL, WOL

Pipe Size	Wall Thickness in Inches				
	0.375" or Less	0.406" to 0.500"	0.562" to 0.688"	0.718" to 0.938"	1.031" to 1.219"
0.5	1.60	2.00	3.80	4.80	5.80
0.75	1.60	2.00	3.80	4.80	5.80
1	1.60	2.00	3.80	4.80	5.80
1.5	1.60	2.00	3.80	4.80	5.80
2	1.60	2.00	3.80	4.80	5.80
2.5	1.75	2.00	4.25	5.75	7.00
3	2.10	2.10	2.40	6.90	6.90
4	2.80	2.80	3.20	6.80	9.20
6	4.20	4.20	4.80	10.20	**13.80**
8	5.60	5.60	6.40	**13.60**	**18.40**
10	7.00	7.00	8.00	**17.00**	**23.00**
12	8.40	8.40	9.60	**20.40**	**27.60**
14	9.80	9.80	11.20	**23.80**	**32.20**
16	11.20	11.20	12.80	**27.20**	**36.80**
18	12.60	12.60	14.40	**30.60**	**41.40**
20	14.00	14.00	16.00	**34.00**	**46.00**
24	16.80	16.80	19.20	**40.80**	**55.20**

SOL, TOL, WOL welds; two times BW (includes cut, prep and fit up).
Wall thickness of the pipe used for weld determines man hours that will apply.
Apply percentages for welding alloys and nonferrous welds.
PWHT butt welds greater than or equal to 0.750".
Preheating and stress-relieving welds are not included.

1.8.4   PWHT Craft Support Labor

Facility—Pump and Compressor Station Piping

Man Hour per Joint

Pipe Size	MH per Joint
0.5	1.00
0.75	1.00
1	1.00
1.5	1.00
2	1.00
2.5	1.63
3	1.95
4	2.60
6	3.90
8	5.20
10	6.50
12	7.80
14	9.10
16	10.40
18	11.70
20	13.00
24	15.60

Insulate and remove insulation from weld joint.

1.8.5   Field Bolt-up of Flanged Joints by Weight Class

Facility—Pump and Compressor Station Piping

Man Hours per Joint

Pipe Size	Pressure Rating		
	150#/300#	600#/900#	1500#/2500#
0.5	0.90	1.20	1.60
0.75	0.90	1.20	1.60
1	0.90	1.20	1.60
1.5	0.90	1.20	1.60
2	0.90	1.20	1.60
2.5	1.13	1.38	1.63
3	1.35	1.65	1.95
4	1.80	2.20	2.60
6	2.70	3.30	3.90
8	3.60	4.40	5.20
10	4.50	5.50	6.50
12	5.40	6.60	7.80
14	6.30	7.70	9.10
16	7.20	8.80	10.40
18	8.10	9.90	11.70
20	9.00	11.00	13.00
24	10.80	13.20	15.60

Man hours per joint to bolt-up valves, expansion joints, flanged fittings, and spools.

1.8.6 Field Handle Valves/Specialty Items by Weight Class

Facility—Pump and Compressor Station Piping

Pipe Size	Man Hours per Valve			Control Valve/Specialty		
	Manual Valve Pressure Rating			Pressure Rating		
	150#/300#	600#/900#	1500#/2500#	150#/300#	600#/900#	1500#/2500#
0.5	0.80	1.30	1.80	1.60	2.60	3.60
0.75	0.80	1.30	1.80	1.60	2.60	3.60
1	0.80	1.30	1.80	1.60	2.60	3.60
1.5	0.80	1.30	1.80	1.60	2.60	3.60
2	0.80	1.30	1.80	1.60	2.60	3.60
2.5	0.88	1.50	2.25	1.75	3.00	3.00
3	1.05	1.80	2.70	2.10	3.60	3.60
4	1.40	2.40	3.60	2.80	4.80	4.80
6	2.10	3.60	5.40	4.20	7.20	7.20
8	2.80	4.80	7.20	5.60	9.60	9.60
10	3.50	6.00	9.00	7.00	12.00	12.00
12	4.20	7.20	10.80	8.40	14.40	14.40
14	4.90	8.40	12.60	9.80	16.80	16.80
16	5.60	9.60	14.40	11.20	19.20	19.20
18	6.30	10.80	16.20	12.60	21.60	21.60
20	7.00	12.00	18.00	14.00	24.00	24.00
24	8.40	14.40	21.60	16.80	28.80	28.80

Man hours only-field handle and erect screwed, flanged, weld end valves, and expansion joints.

1.9 Schedule G—Alloy and Nonferrous Weld Factors

Welding Percentages for Alloy and Nonferrous Metals

Material Classification Group Numbers

Pipe Size	1	2	3	4	5	6	7	8
2	0.25	0.54	0.20	0.58	2.11	2.25	0.225	0.45
3	0.275	0.58	0.23	0.61	2.15	2.32	0.25	0.495
4	0.30	0.61	0.25	0.68	2.22	2.35	0.28	0.54
5	0.315	0.63						0.57
6	0.345	0.65	0.30	0.75	2.28	2.40	0.30	0.62
8	0.39	0.74	0.50	0.88	2.38	2.50	0.34	0.70
10	0.425	0.85	0.75	0.95	2.45	2.75	0.375	0.765
12	0.45	2.00	0.80	2.04	2.50	3.00	0.40	0.81
14	0.49	2.15						0.88
16	0.525	2.23						0.945
18	0.59	2.30						2.06
20	0.65	2.45						2.17
24	0.73							2.24

Group 1—chrome molybdenum steel, chrome—12%–13%, moly—to 1%.
Group 2—18-8 stainless steel, TY, 304, 316, 347.
Group 3—copper, brass, everdur.
Group 4—aluminum, monel, and copper, chrome–nickel.
Group 5—nickel.
Group 6—hastelloy.
Group 7—galvanized.
Group 8—A335—P91.

1.10   Schedule H—Industrial Plant Piping

Standard Labor Estimating Units		
Facility—Industrial Plant	**Large Bore Piping**	**Small Bore Piping**
	Unit of Measure	Unit of Measure
Description	Man Hours per Unit	Man Hours per Unit
PVC Pipe	Diameter Inch	MH/EA
Handle and install PVC pipe	0.046	0.12
PVC solvent joints	0.15	0.35
Copper Pipe		
Handle and install copper pipe	0.10	0.18
Copper sweat joints	0.70	1.40
Victaulic Pipe		
Victaulic couplings	0.35	0.70
Victaulic grooving	0.35	0.70
Mechanical couplings (dresser)	0.50	1.00
Screwed Pipe		
Handle and install screwed pipe	0.035	
Screwed joints (measure, cut, thread, makeup)	0.30	
Miscellaneous		
Hydro test	0.12×pipe handle	
In-line instruments (PIs, TIs)		1.20
Socketweld		Per SW table

1.10.1 Welding, Carbon Steel, Socketweld

Facility—Industrial Plant

Man Hour per SW

Pipe Size	Socket Welds	
Inches	**Schedule 40 and 80**	**Schedule 100 and Greater**
0.5	1.2	1.3
0.75	1.2	1.3
1	1.3	1.4
1.5	1.4	1.6
2	1.5	1.8

Includes place and fit up.

1.10.2 Hydrostatic Testing

Facility—Industrial Plant

Man Hours per Lineal Foot

Pipe Size	Wall Thickness in Inches				
	0.375″ or Less	**0.406″ to 0.500″**	**0.562″ to 0.688″**	**0.718″ to 0.938″**	**1.031″ to 1.219″**
0.5	0.022	0.028	0.034	0.042	0.060
0.75	0.022	0.028	0.034	0.042	0.060
1	0.022	0.028	0.034	0.042	0.060
1.5	0.022	0.028	0.034	0.042	0.060
2	0.022	0.028	0.034	0.042	0.060
2.5	0.021	0.027	0.033	0.042	0.060
3	0.025	0.032	0.040	0.050	0.072
4	0.034	0.043	0.053	0.067	0.096
6	0.050	0.065	0.079	0.101	0.144
8	0.067	0.086	0.106	0.134	0.192
10	0.084	0.108	0.132	0.168	0.240
12	0.101	0.130	0.158	0.202	0.288
14	0.118	0.151	0.185	0.235	0.336
16	0.134	0.173	0.211	0.269	0.384
18	0.151	0.194	0.238	0.302	0.432
20	0.168	0.216	0.264	0.336	0.480
24	0.202	0.259	0.317	0.403	0.576

Man hours to place/remove blinds, open/close valves, removal/replacement of valves and specialty items and pipe sections as required, and drain lines after testing.

1.10.3   Cut and Prep Weld Joint (Industrial Piping)

Facility—Industrial Plant

Man Hours per Joint

Pipe Size	Wall Thickness in Inches				
	0.375″ or Less	0.406″ to 0.500″	0.562″ to 0.688″	0.718″ to 0.938″	1.031″ to 1.219″
0.5	0.08	0.09	0.10	0.12	0.14
0.75	0.11	0.14	0.15	0.17	0.20
1	0.15	0.18	0.20	0.23	0.27
1.5	0.23	0.27	0.30	0.35	0.41
2	0.30	0.36	0.40	0.46	0.54
2.5	0.38	0.45	0.50	0.58	0.68
3	0.45	0.54	0.60	0.69	0.81
4	0.60	0.72	0.80	0.92	1.08
6	0.90	1.08	1.20	1.38	1.62
8	1.20	1.44	1.60	1.84	2.16
10	1.50	1.80	2.00	2.30	2.70
12	1.80	2.16	2.40	2.76	3.24
14	2.10	2.52	2.80	3.22	3.78
16	2.40	2.88	3.20	3.68	4.32
18	2.70	3.24	3.60	4.14	4.86
20	3.00	3.60	4.00	4.60	5.40
24	3.60	4.32	4.80	5.52	6.48

1.10.4 PVC Piping

Facility—Industrial Plant

Pipe Size	Handle Pipe MH/LF	Solvent Joint MH/Joint	Valve Handling 150#/300#	CV/SP	Bolt-up MH/Joint 150#/300#
0.5	0.12	0.35	0.40	0.80	1.00
0.75	0.12	0.35	0.40	0.80	1.00
1	0.12	0.35	0.40	0.80	1.00
1.5	0.12	0.35	0.40	0.80	1.00
2	0.12	0.35	0.40	0.80	1.00
2.5	0.00	0.38	0.63	1.25	1.00
3	0.00	0.45	0.75	1.50	1.20
4	0.00	0.60	1.00	2.00	1.60
6	0.00	0.90	1.50	3.00	2.40
8	0.00	1.20	2.00	4.00	3.20
10	0.00	1.50	2.50	5.00	4.00
12	0.00	1.80	3.00	6.00	4.80

Erect pipe man hours include all labor to unload, store in lay down, haul, rig, and align in place.
Solvent joints include cut, square, ream, fit up and make joint.
Handle valves man hours include field handling and erection of valves.
Man hours per joint to bolt-up valves.

1.10.5 Copper Piping

Facility—Industrial Plant

Pipe Size	Pipe MH/LF	Handle Sweat Joint MH/Joint	Bolt-up 150#/300#	600#/900#	CV/SP	Bolt-up 150#/300#	600#/900#
0.5	0.18	1.40	0.45	0.90	1.80	1.00	1.30
0.8	0.18	1.40	0.45	0.90	1.80	1.00	1.30
1	0.18	1.40	0.45	0.90	1.80	1.00	1.30
1.5	0.18	1.40	0.75	1.50	3.00	1.00	1.30
2	0.18	1.40	0.90	1.80	3.60	1.00	1.30
2.5	0.25	1.75	1.13	2.25	4.50	1.00	1.50
3	0.30	2.10	1.35	2.70	5.40	1.20	1.80
4	0.40	2.80	1.80	3.60	7.20	1.60	2.40
6	0.60	4.20	2.70	5.40	10.80	2.40	3.60
8	0.80	5.60	3.60	7.20	14.40	3.20	4.80
10	1.00	7.00	4.50	9.00	18.00	4.00	6.00
12	1.20	8.40	5.40	10.80	21.60	4.80	7.20

Erect pipe man hours include all labor to unload, store in lay down, haul, rig, and align in place.
Man hours include handling and jointing or making on solder-type brass or copper joints.
Handle valves man hours include field handling and erection of valves.
Man hours per joint to bolt-up valves.

1.10.6 Screwed Pipe

Facility—Industrial Plant

Pipe Size	Handle Pipe	Screwed Joint
Inches	**MH/LF**	**MH/Joint**
0.5	0.10	0.30
0.75	0.10	0.30
1	0.10	0.30
1.5	0.10	0.45
2	0.10	0.60

Handle and install screwed pipe.
Screwed joints (measure, cut, thread, makeup).

1.11 Schedule I—Underground Drainage Piping for Industrial Plants

Standard Labor Estimating Units	
Facility—Industrial Plant (Underground Drainage Piping)	
	Unit of Measure
	Man Hours per Unit
Description	**Diameter Inch Feet**
Cast iron—lead and mechanical joint	
Handle and install pipe	0.03
Handle and install fittings	3′ X fitting
	MH/Diameter Inch
Lead and mechanical joints	0.20
	Diameter Inch Feet
Cast iron soil—no hub	
Handle and install pipe	0.025
Handle and install fittings	3′ X fitting
	MH/Diameter Inch
Couplings	0.10
Pipe trench	MH/CY
Trench excavation—backhoe 1 CY bucket	0.15
Trench excavation—backhoe 3/4 CY bucket	0.20
Trench excavation—backhoe 1/2 CY bucket	0.25
Trench excavation—hand (to 4′ deep)	2.40
Backfill and compaction—loader and compactors	0.60
Backfill and compaction—hand and compactors	3.40
Sand bedding and shading W/loader	0.40
Shoring (Place/Remove)	**MH/SF**
Laborer	5.50
Carpenter	3.00
	Diameter Inch Feet
Vitrified clay pipe	0.02
Make-ons	0.08
Concrete pipe	0.018
Cement poured joint	0.06

1.11.1   Handle and Install Pipe, Cast Iron—Lead and Mechanical Joint

Facility—Industrial Plant (Underground Drainage Piping)

Pipe Size	Pipe Set Align	Lead and Mechanical Joint	Handle Install Fitting
Inches	MH/LF	MH/JT	MH/EA
4	0.12	0.80	0.24
6	0.18	1.20	0.36
8	0.24	1.60	0.48
10	0.30	2.00	0.60
12	0.36	2.40	0.72
12	0.36	2.40	0.72
16	0.48	3.20	0.96
18	0.54	3.60	1.08
20	0.60	4.00	1.20
24	0.72	4.80	1.44

Pipe man hours include handle, haul, place, and align in trench.

1.11.2   Handle and Install Pipe, Cast Iron Soil—No Hub

Facility—Industrial Plant (Underground Drainage Piping)

Pipe Size	Pipe Set and Align	Handle Install Fitting	Couplings
Inches	MH/LF	MH/EA	MH/EA
4	0.10	0.24	0.40
6	0.15	0.36	0.60
8	0.20	0.48	0.80
10	0.25	0.60	1.00
12	0.30	0.72	1.20
12	0.30	0.72	1.20
16	0.40	0.96	1.60
18	0.45	1.08	1.80
20	0.50	1.20	2.00
24	0.60	1.44	2.40

Pipe man hours include handle, haul, place, and align in trench.

1.11.3   Handle and Install Pipe, Concrete and Vitrified Clay

Facility—Industrial Plant (Underground Drainage Piping)

	Vitrified Clay Pipe		Concrete Pipe	
	Pipe Set		Pipe Set	Cement
Pipe Size	Align	Make-On	Align	Poured Joint
Inches	MH/LF	MH/JT	MH/LF	MH/JT
4	0.08	0.32	0.07	0.24
6	0.12	0.48	0.11	0.36
8	0.16	0.64	0.14	0.48
10	0.20	0.80	0.18	0.60
12	0.24	0.96	0.22	0.72
12	0.24	0.96	0.22	0.72
16	0.32	1.28	0.29	0.96
18	0.36	1.44	0.32	1.08
20	0.40	1.60	0.36	1.20
24	0.48	1.92	0.43	1.44
30	0.60	2.40	0.54	1.80
36	0.72	2.88	0.65	2.16
42	0.84	3.36	0.76	2.52
48	0.96	3.84	0.86	2.88
60	1.20	4.80	1.08	3.60

Pipe man hours include handle, haul, place, and align in trench.

1.11.4   Pipe Trench-Machine, Hand Excavation, Backfill, and Sand Bedding

Pipe Trench-Machine and Hand Excavation	
Facility—Industrial Plant (Underground Drainage Piping)	
Machine Excavation	MH/CY
Trench excavation—backhoe 1 CY bucket	0.15
Trench excavation—backhoe 3/4 CY bucket	0.20
Trench excavation—backhoe 1/2 CY bucket	0.25
Hand excavation	
Trench excavation—hand (to 4′ deep)	2.40
Pipe Trench-Backfill and Compaction-Loader and Compactors and Hand and Compactors	
Facility—Industrial Plant (Underground Drainage Piping)	
Backfill and compaction	MH/CY
Backfill and compaction—loader and compactors	0.60
Backfill and compaction—hand and compactors	3.40
Pipe Trench-Sand Bedding and Shading with Loader	
Facility—Industrial Plant (Underground Drainage Piping)	
	MH/CY
Sand bedding and shading W/loader	0.40
Pipe Trench-Shoring and Bracing Trenches	
Facility—Industrial Plant (Underground Drainage Piping)	
Shoring (Place/Remove)	MH/SF
Laborer	5.50
Carpenter	3.00

1.12   Schedule J—Simple Foundations for Industrial Plants

	MH/CY	MH/SF	MH/LF	MH/lb	MH/diameters in feet
Standard Labor Estimating Units					
Facility—Industrial Plant (Simple Foundation)					
Structure excavation—backhoe	0.20				
Structure excavation—hand	2.80				
Structure backfill and compact—loader and wacker	0.60				
Structure backfill and compact—hand	3.00				
Edge forms—slabs and foundations			0.10		
Fabricate, install, strip foundation forms—1 use		0.30			
Fabricate, install, strip pedestal forms—1 use		0.25			
Fabricate, install, strip wall forms—1 use		0.20			
Fabricate and install reinforcing steel				0.03	
Layout templates and set anchor bolts					0.25
Set embedded steel—curb angle, ETC				0.04	
Place concrete—from truck below grade	0.80				
Place concrete—slabs at grade	1.00				
Place concrete—pedestals and walls	2.00				
Finish flat concrete		0.12			
Patch and sack concrete		0.05			
Install mesh			0.02		

1.13   Schedule K—Pipe Supports and Hangers

Facility—Industrial Plant-Pipe Supports and Hangers

Pipe Support Spacing	Span in Feet	
Pipe Size	**Water**	**Steam**
Inches	**Service**	**Gas and Air**
1	7	9
1.5	9	12
2	10	13
2.5	11	14
3	12	15
3.5	13	16
4	14	17
5	16	19
6	17	21
8	19	24
10	22	28
12	23	30
14	25	32
16	27	35
18	28	37
20	30	39
24	32	42
Standard Labor Estimating Units		
Type of Hanger	MH	
Pipe size (1″ through 4″)		
Clevis, band, ring, and expansion hangers	3.8	
Trapeze hanger	6.0	
Pipe size		
6″	5.0	
8″	5.3	
10″	5.6	
12″	6.0	
Shoes and guides	2.0	
Engineered supports; spring, sway, snubbers	1.0 MH/DI	
Structural supports (angles, channels, beams, tube steel)—0.04 per lb		

1.14   Standard and Line Pipe-Wall Thickness

Pipe Size	40	STD	60	80	XH	100	120	140	160	XXH
0.5	0.109	0.109		0.147	0.147				0.188	0.294
0.75	0.113	0.113		0.154	0.154				0.219	0.308
1	0.133	0.133		0.179	0.179				0.250	0.358
1.5	0.145	0.145		0.200	0.200				0.281	0.400
2	0.154	0.154		0.218	0.218				0.344	0.436
2.5	0.203	0.203		0.276	0.276				0.375	0.552
3	0.216	0.216		0.300	0.300				0.438	0.600
4	0.237	0.237		0.337	0.337		0.438		0.531	0.674
6	0.280	0.280		0.432	0.432		0.562		0.719	0.864
8	0.322	0.322	0.406	0.500	0.500	0.594	0.719	0.812	0.906	0.875
10	0.365	0.365	0.500	0.594	0.500	0.719	0.844	1.000	1.125	1.000
12	0.406	0.375	0.562	0.688	0.500	0.844	1.000	1.125	1.312	1.000
14	0.438	0.375	0.594	0.750	0.500	0.938	1.094	1.250	1.406	
16	0.500	0.375	0.656	0.844	0.500	1.031	1.219	1.438	1.594	
18	0.562	0.375	0.750	0.938	0.500	1.156	1.375	1.562	1.781	
20	0.594	0.375	0.812	1.031	0.500	1.281	1.500	1.750	1.969	
22		0.375	0.875	1.125	0.500	1.375	1.625	1.875	2.125	
24	0.688	0.375	0.969	1.219	0.500	1.531	1.812	2.062	2.344	

Combined Cycle Power Plant Equipment

2

2.1 Scope of Field Work Required for Each F Class CTG

F Class CTG

Complete Scope on One Page

Set, align, key, and couple centerline equipment
Foundation bolting arrangement
Set gas turbine
Install load coupling to turbine (inlet end)
Set generator
Collector assembly and turning gear
Install turbine exhaust diffuser
Alignment of centerline equipment
Generator turbine alignment

Set modules and skids
Set accessory package
Set liquid fuel and atomizing air skid
Set packaged electrical electronic control compartment
Water injection skid
Fire protection skid
Set water cooling tower or module
Set load commutated inverter (LCI) excitation compartment
Set DC link reactor (static start)
Set isolation/excitation transformer (static start)
Set bus accessory compartment (BAC)
Set liquid fuel forwarding skid
Air processing skid
Set water wash skid
Set cooling fan module
Set static start excitation substation
Main power transformer
Set and assemble ISO phase duct

Industrial Piping and Equipment Estimating Manual. http://dx.doi.org/10.1016/B978-0-12-813946-2.00002-2

Place other equipment

     Set hydrogen dryer and purge assembly (GEN)

     Set liquid detectors (GEN)

     Install batteries

     Install (PEECC) air conditioners

     Set drain collection tank (BOP)

Enclosures and Barriers

     SET generator terminal enclosure (GEN)

     Install turbine enclosure grading and walkways

     Install turbine load compartment

     Install enclosure to base

Assemble air inlet system

     Erect turbine air inlet plenum

     Erect turbine air inlet ductwork

     Erect turbine air inlet filter house

     Erect turbine exhaust system/stack

2.2 F Class CTG Estimating Data

2.2.1 Centerline Equipment, Modules/Skids, and Equipment Sheet 1

Description	MH	Unit
Centerline Equipment		
Set fixators	3.50	EA
Grout fixators	9.20	EA
Load coupling, collector, and turning gear	9.25	ton
Exhaust diffusers	9.25	ton
Rough alignment	1.42	ton
Final alignment	2.10	ton
Turbine and generator	0.50	ton
Weld radiation shields	1.35	EA
Modules and Skids		
Weight <20,000 lb	5.2	ton
20,000 lb > weight < 60,000 lb	4.8	ton
60,000 lb > weight <= 100,000 lb	3.2	ton
Generator terminal enclosure	6.20	ton
Place Equipment		
Set hydrogen dryer and purge assembly	30.00	EA
Set liquid detectors	20.00	Lot
Install batteries (pallet #1, 2, 3)	20.00	ton
Install air conditioners	42.00	Lot
Set drain collection tank	30.00	EA
Set load excitation compartment	20.00	EA
Set DC link reactor	20.00	EA
Set isolation/excitation transformer	20.00	EA

2.2.2 Generator Enclosure, Enclosure, Grating, Walkways, and Compartment Sheet 2

Description	MH	Unit
Generator Terminal Enclosure	14.40	ton
Turbine Enclosure, Grating, and Walkways		
Roof section, side, end, and frame upper/lower	12.40	ton
Door	10.00	EA
Panels and floor unit	24.00	ton
Make-up field joints	0.75	LF
Support steel	64.00	ton
Turbine Load Compartment		
Left/right lower, top half section	60.00	ton
Bottom plate	60.00	ton
Door	10.00	EA
Field joints	0.75	LF
Enclosure	280.00	EA

2.2.3   Support Steel, Inlet Filter House, Transition Duct, Inlet Duct, Plenum Sheet 3

Description	MH	Unit
Air Filter System		
Support steel	14.50	ton
Inlet Filter House		
Sloped roof	8.40	ton
Filter modules	8.40	ton
Walkway assembly	8.40	ton
D15 weather hoods	8.40	ton
Hopper	20.00	EA
Pipe/tubing	1.94	LF
Hand rails	0.25	LF
Filter cartridges	0.75	EA
Field joints	0.75	LF
Inlet Filter House Transition Duct		
Duct sections	28.00	ton
Make-up duct field joints	0.75	LF
Inlet Duct Work		
Transition, elbow upper/lower duct	8.40	ton
Heating, silencer w/panels, empty duct	8.40	ton
Expansion joint (kit)	40.00	kit
Make-up duct field joints	0.75	LF
Inlet Plenum		
Plenum support structure	40.00	ton
Plenum extension and cone section	28.00	ton
Lower/upper quarter section	28.00	ton
Lot 1, 2, and 3	40.00	ton
Make-up duct field joints	0.75	LF

2.2.4   Exhaust Duct and Stack Sheet 4

Description	MH	Unit
Exhaust System		
Exhaust Duct		
Upstream and downstream Upper/lower quadrant	8.40	ton
Downstream expansion joint (kit)	40.00	EA
Support steel (kit)	40.00	kit
Make-up duct field joints	0.75	LF
Exhaust Duct/Stack		
Upstream and downstream duct transition	8.40	ton
Duct silencer	8.40	ton
Lower and upper stack transition	8.40	ton
Stack segments	8.40	ton
Silencer	5.20	ton
Support steel	24.00	ton
Make-up field joints	0.75	LF

2.3   F Class CTG Installation Estimate

2.3.1   Centerline Equipment, Modules/Skids, and Equipment Sheet 1

Estimate—Centerline Equipment		Actual		Estimate				
Description	MH	Quantity	Unit	Quantity	Unit	BM	IW	MW
Centerline Equipment						**0**	**0**	**0**
Set fixators	3.5	16.0	EA	0.0	EA			0
Grout fixators	9.2	16.0	EA	0.0	EA			0
Rig, lift ,and set turbine down onto fixators	0.5	188.5	ton	0.0	ton		0	
Install load coupling to turbine	9.25	2.5	ton	0.0	ton			0
Rig, lift, and set generator	0.5	280.0	ton	0.0	ton		0	
Install collector assembly	9.25	10.0	ton	0.0	ton			0
Weld radiation shields and install thermocouple	1.35	27.0	EA	0.0	EA		0	
Rig and set exhaust diffuser	9.25	14.1	ton	0.0	ton			0
Rough alignment, centerline equipment	1.42	495.1	ton	0.0	ton			0
Generator–turbine final alignment	2.1	495.1	ton	0.0	ton			0

Estimate—Modules and Skids		Actual		Estimate				
Modules and Skids						**0**	**0**	**0**
Set accessory package	3.20	37.5	ton	0.0	ton			0
Set liquid fuel and atomizing air skid	4.80	30.0	ton	0.0	ton			0
Set electric control compartment	4.80	23.0	ton	0.0	ton			0
Water injection skid	4.80	15.0	ton	0.0	ton			0
Fire protection skid	5.20	10.0	ton	0.0	ton			0
Set water cooling module	3.20	49.8	ton	0.0	ton			0
Set bus accessory compartment	5.20	7.5	ton	0.0	ton			0
Set liquid fuel forwarding skid	5.20	5.9	ton	0.0	ton			0
Air processing skid	5.20	1.3	ton	0.0	ton			0
Set water wash skid	4.80	10.9	EA	0.0	EA			0
Set cooling fan module	5.20	5.5	Lot	0.0	Lot			0

Estimate—Place Equipment		Actual		Estimate				
Place Equipment						**0**	**0**	**0**
Set hydrogen dryer and purge assembly	30.0	1.0	Lot	0.0	Lot			0
Set liquid detectors	20.0	1.0	EA	0.0	EA			0
Install batteries (pallet #1, 2, 3)	20.0	1.0	EA	0.0	EA			0
Install air conditioners	42.0	1.0	EA	0.0	EA			0
Set drain collection tank	30.0	1.0	EA	0.0	EA			0
Set load excitation compartment	20.0	1.0	EA	0.0	EA			0
Set DC link reactor	20.0	1.0	EA	0.0	EA			0
Set isolation/excitation transformer	20.0	1.0	EA	0.0	EA			0

2.3.2 Generator Enclosure, Enclosure, Grating, Walkways, and Compartment Sheet 2

Estimate Enclosures and Barriers	Historical			Estimate				
Description	MH	Quantity	Unit	Quantity	Unit	BM	IW	MW
Enclosures and Barriers **Generator Terminal Enclosure**						0	0	0
Set generator terminal enclosure	14.4	12.5	ton	0.0	ton		0	

Estimate—Enclosure, Grating and Walkways	Historical			Estimate				
Turbine Enclosure, Grating, and Walkways						0	0	0
Roof section	12.4	5.5	ton	0.0	ton		0	
Side upper; frame end upper	12.4	6.5	ton	0.0	ton		0	
Door	10.0	4.0	EA	0.0	EA		0	
Frame side/end upper/lower	12.4	16.9	ton	0.0	ton		0	
Panels	24.0	9.2	ton	0.0	ton		0	
Make-up field joints; roof, frames and panels	0.75	723.6	LF	0.0	LF		0	
Floor unit	24.0	5.3	ton	0.0	ton		0	
Support steel	64.0	0.4	ton	0.0	ton		0	
Make-up field joints; floor units	0.75	773.0	LF	0.0	LF		0	

Estimate—Turbine Load Compartment	Historical			Estimate				
Turbine Load Compartment						0	0	0
Left lower quarter section	60.0	0.4	ton	0.0	ton		0	
Right lower quarter section	60.0	0.4	ton	0.0	ton		0	
Top half section	60.0	0.7	ton	0.0	ton		0	
Bottom plate	60.0	0.5	ton	0.0	ton		0	
Doors	10.0	2.0	EA	0.0	EA		0	
Field joints	0.8	72.6	LF	0.0	LF		0	
Install Enclosure to Base						0	0	0
Set enclosure on base	280.0	1.0	EA	0.0	EA		0	

Estimate—Inlet Filter House	Historical			Estimate				
Description	MH	Quantity	Unit	Quantity	Unit	BM	IW	MW
Air Filter System								
Inlet filter house						0	0	0
Sloped roof	8.40	2.1	ton	0.0	ton	0		
Hopper	20.00	1.0	ton	0.0	ton	0		
Pipe/tubing	1.94	49.7	LF	0.0	LF	0		
Hand rails	0.25	21.1	LF	0.0	LF	0		
Filter cartridges	0.75	1020.0	EA	0.0	EA	0		
Filter modules (center)	8.40	15.5	ton	0.0	ton	0		
Filter modules (left and right hand)	8.40	9.3	ton	0.0	ton	0		
Walkway assembly (center)	8.40	4.7	ton	0.0	ton	0		
Walkway assembly (LH and RH)	8.40	3.5	ton	0.0	ton	0		
D15 weather hoods	8.40	9.9	ton	0.0	ton	0		
Field joints	0.75	620.6	LF	0.0	LF	0		

Estimate—Inlet Filter House Transition Duct	Historical			Estimate				
Inlet Filter House Transition Duct						0	0	0
Top section	28.00	1.3	ton	0.0	ton	0		
Floor section	28.00	1.5	ton	0.0	ton	0		
Side section	28.00	1.0	ton	0.0	ton	0		
Side section	28.00	0.6	ton	0.0	ton	0		
Bell section	28.00	1.3	ton	0.0	ton	0		
Make-up duct field joints	0.75	235.4	LF	0.0	LF	0		

Estimate—Support Steel	Historical			Estimate				
Support Steel						0	0	0
Erect support steel	14.50	86.0	ton	0.0	ton	0		

Estimate—Inlet Duct Work	Historical			Estimate				
Inlet Duct Work						0	0	0
Transition duct	8.40	5.9	ton	0.0	ton	0		
Expansion joint (kit)	40.00	1.0	kit	0.0	kit	0		
Elbow duct upper w/trash screen	8.40	8.7	ton	0.0	ton	0		
Elbow duct lower	8.40	10.9	ton	0.0	ton	0		
Heating duct	8.40	4.8	ton	0.0	ton	0		
Silencer duct w/panels	8.40	22.3	ton	0.0	ton	0		
Empty duct	8.40	3.1	ton	0.0	ton	0		
Make-up duct field joints	0.75	1072.2	LF	0.0	LF	0		

Estimate—Inlet Plenum	Historical			Estimate				
Inlet Plenum						0	0	0
Plenum support structure	40.00	**2.2**	ton	0.0	ton	0		
Plenum extension	28.00	**0.8**	ton	0.0	ton	0		
Plenum cone (inlet case ext)	28.00	**1.6**	ton	0.0	ton	0		
Lower/upper quarter section	28.00	**14.9**	ton	0.0	ton	0		
Lot 1, 2, and 3	40.00	**2.5**	ton	0.0	ton	0		
Make-up duct field joints	0.75	**227.5**	LF	0.0	LF	0		

2.3.4 Exhaust Duct and Stack Sheet 4

Estimate—Exhaust Duct		Historical		Estimate				
Description	MH	Quantity	Unit	Quantity	Unit	BM	IW	MW
Exhaust System								
Exhaust Duct						0	0	0
Upstream upper/lower quadrant	8.40	11.4	ton	0.0	ton	0		
Downstream upper/lower quadrant	8.40	10.3	ton	0.0	ton	0		
Downstream expansion joint (kit)	40.00	1.0	EA	0.0	EA	0		
Support steel (kit)	40.00	1.0	kit	0.0	kit	0		
Make-up duct field joints	0.75	148.8	LF	0.0	LF	0		

Estimate—Exhaust Duct/ Stack		Historical		Estimate				
Exhaust Duct/Stack						0	0	0
Upstream duct transition	8.40	12.0	ton	0.0	ton	0		
Duct silencer	8.40	28.0	ton	0.0	ton	0		
Downstream duct transition	8.40	6.0	ton	0.0	ton	0		
Lower stack transition	8.40	4.5	ton	0.0	ton	0		
Silencer	5.20	24.0	ton	0.0	ton	0		
Upper stack transition	8.40	4.5	ton	0.0	ton	0		
Stack segments	8.40	16.0	ton	0.0	ton	0		
Make-up duct field joints	0.75	687.4	LF	0.0	LF	0		
Support steel	24.00	6.1	ton	0.0	ton	0		

2.4 F Class CTG-Equipment Installation Man Hours

Facility-Combined Cycle Power Plant	Actual	Estimate			
Description	MH	BM	IW	MW	MH
Centerline equipment	2463	0	0	0	0
Modules and skids	1047	0	0	0	0
Place equipment	202	0	0	0	0
Generator terminal enclosure	180	0	0	0	0
Turbine enclosure, grating and walkways	1891	0	0	0	0
Turbine load compartment	199	0	0	0	0
Inlet filter house	1730	0	0	0	0
Inlet filter house transition duct	337	0	0	0	0
Support steel	1247	0	0	0	0
Inlet duct work	1311	0	0	0	0
Inlet plenum	841	0	0	0	0
Exhaust duct	374	0	0	0	0
Exhaust duct/stack	1382	0	0	0	0
Equipment installation man hours	**13202**	**0**	**0**	**0**	**0**

2.5   General Scope of Field Work Required for Each Reheat Double Flow STG Sheet 1

Reheat Double Flow STG
Scope of Work-Field Erection

Foundation preparation and soleplate setting
> Set and grout foundation support system
> Grout encapsulated support system
> Turbine, front standard baseplate, and generator soleplates

Centerline installation
Low pressure turbine installation
> Set lower exhaust hood
> Install lower inner casing and packing casings
> Install and assemble upper inner casing
> Install upper half exhaust hood
> Close exhaust hood horizontal joint and check foot contact
> Exhaust hood centerline anchor/key block soleplate
> Preliminary tops-on exhaust hood alignment
> Turning gear standard to exhaust casing alignment
> Weld exhaust hood to condenser
> Final grout exhaust hood and turning gear standard soleplates
> Final tops-on alignment
> Fit exhaust hood keys
> Set and fit bearing rings

HP/IP turbine and front standard installation
> Set front standard baseplate
> Front standard baseplate checks
> Set the HP/IP turbine assembly in position
> Front standard to baseplate checks

Final turbine assembly and alignment
> Remove upper exhaust hood and upper inner casing
> Remove upper–inner casing
> Final tops-off alignment
> Install and align lower half low pressure diaphragms
> Install and fit No. 3 and 4 bearings
> Install low pressure rotor
> Install low pressure oil deflectors
> Install upper half low pressure bearings and bearing rings
> Measure and record internal low pressure turbine clearances
> Install upper half low pressure diaphragms
> Final installation of upper inner casing
> Final installation and assembly of upper half exhaust hood
> HP/IP to LP assembly

Turbine–generator acoustic enclosure
> Grout the front standard baseplate
> HOM drop check
> Shell ARM load checks
> HP/IP connections
> Final alignment of HP/IP rotor to LP rotor

Assemble "A" coupling (HP/IP and LP rotor coupling)
Install HP/IP and mid-standard elevation and centerline keys
Install crossover
Install and align turning gear
Install heat retention insulation
Turbine-generator acoustic enclosure

Lube oil system installation

Lube oil tank
Lube oil tank filling
Lube oil system flushing

Hydraulic power oil system installation

Install the hydraulic oil power system
Hydraulic power oil system arrangement
Hydraulic power unit (HPU) outline and field connections
Hydraulic power system flush

Generator installation

Set generator terminal compartment
Rigging and moving generator
Set and rough align generator stator
Final generator alignment
Assemble LP turbine–generator coupling

Generator electrical systems

2.6 Double Flow STG Estimating Data

2.6.1 Centerline Equipment Sheet 1

Description	MH	Unit
Centerline equipment		SF
Layout foundations	0.12	SF
Clean protective coating	0.25	LF
Weld LO supply and drain downcomers	6.00	EA
Bolt connections	0.50	EA
Install and grout retaining strips	0.50	LF
Set lower turbine and exhaust hood	1.02	ton
Clean and set turning gear standard	80.00	EA
Move exhaust hood	80.00	EA
Rough alignment	1.42	ton
Seal weld mid-standard	0.50	LF
Rig and install inner casing	2.56	ton
Install lower half packing casing	20.00	EA
Rig and install upper casing	2.56	ton
Final alignment	2.10	ton
Install upper exhaust hood halves	2.56	ton
Rough alignment	2.56	ton
Crossover flange level check	8.00	EA
Adjust left and right side anchor blocks	20.00	EA
Weld anchor blocks exhaust hood	10.00	EA
Weld exhaust hood condenser neck	1.50	LF
Alignment during welding	2.10	ton
Install condenser joint shield	40.00	EA
Clean area and install grout sole plates	16.00	EA
Fit internal exhaust hood keys	1.00	EA
Set T2, T3, and T4 bearing rings	10.00	EA

2.6.2 HP/IP Turbine and Front Standard Sheet 2

Description	MH	Unit
Set fixators	3.50	EA
Clean and set front standard plate	20.00	EA
Rough alignment	1.42	ton
Final alignment	2.10	ton
Set assembled HP/IP turbine	1.70	ton
Remove turbine end shipping brace	4.00	EA

2.6.3 Final Turbine Assembly and Alignment Sheet 3

Description	MH	Unit
Final Turbine Assembly and Alignment		
Remove dowels and bolts from exhaust hood	0.55	EA
Rig and remove upper exhaust hood	1.56	ton
Loosen and remove dowels and bolts	0.50	EA
Rig and remove upper inner casing	2.56	ton
Alignment (T3 and T4 inboard oil deflector)	2.10	ton
Clean protective coating	0.25	LF
Rig and install diaphragms	14.00	EA
Rough alignment	1.42	ton
Bearings, bearing rings, and standard fits	0.25	LF
Install bearings	16.00	EA
Install upper half bearing rings	16.00	EA
Rig and install low pressure rotor	1.65	ton
Install lower half oil deflectors	10.00	EA
Install upper half bearings and rings	20.00	EA
Rough alignment	0.04	lb
Rig and install the upper inner casing	2.56	ton
Install horizontal joint shields	0.50	LF
Final alignment	2.10	ton
Rig and install the upper exhaust hood	1.56	ton
Bolt connection-vertical joint	0.50	EA
Install atmosphere relief diaphragm	16.00	EA
Set T2 lower half bearings	20.00	EA
Rig and lift generator end of rotor	1.60	ton
Remove shipping brackets	4.00	EA
Install upper T1 and T2 bearings	20.00	EA

2.6.4 Turbine–Generator Acoustic Enclosure Sheet 4

Description	MH	Unit
Generator Acoustic Enclosure		
Grout front standard base plate	1.40	SF
Install/remove generator end rotor support	40.00	EA
Final alignment	2.10	ton
Clean coupling and associated hardware	0.25	LF
Assemble, fit, and dowel "A" coupling guard	30.00	EA
Install and fit "A" coupling covers	10.00	EA
Install and fit upper half oil deflectors	10.00	EA
Machine and install center keys	2.58	EA
Rough alignment	1.42	ton
Remove shipping skid and coatings	0.25	LF
Remove expansion joint locking device	20.00	EA
54″ diameter fit up and weld-HP/LP turbine	1.15	DI
54″ diameter fit up and weld-assembly	1.15	DI
Install 4″diameter crossover pipe	3.64	LF
54″ diameter flange bolt up	0.45	DI
Install turning gear to rotor	40.00	EA
Final alignment	2.10	ton
Dowel turning gear to standard	1.20	EA
Left, right, middle, and end turbine sections	40.00	EA
Turbine bottom plate	40.00	EA
Doors	8.00	EA
Field joints	16.00	EA

2.6.5 Generator Installation, Lube Oil System, Hydraulic Power Oil System Sheet 5

Description	MH	Unit
Generator Installation		
Assemble cribbing and set compartment	120.00	EA
Clean and install grunions	10.00	EA
Rig, lift, and move and set generator	1.39	ton
Remove and install generator outer shields	20.00	EA
Install generator journal bearings	10.00	EA
Bolt-up outer end shield and dowel	1.00	EA
Install hydrogen seal rings	25.00	EA
Install and align oil deflectors	10.00	EA
Set and rough alignment-stator	1.42	ton
Final alignment	2.10	ton
Grout sole plate	16.00	EA
Install studs and align coupling guard	40.00	EA
Lube Oil System Installation		
Set lube oil tank	60.00	EA
Align lube oil pumps	20.00	EA
Install lube oil piping	2.20	LF
Install lube oil conditioner	60.00	EA
Hydraulic Power Oil System Installation		
Set hydraulic power unit	80.00	EA
Align hydraulic power pumps	20.00	EA
Install hydraulic power piping	2.20	LF

2.7 Double Flow STG Installation Estimate

2.7.1 Centerline Equipment t Sheet 1

Estimate—Centerline Equipment	Actual			Estimate				
Description	MH	Quantity	Unit	Quantity	Unit	BM	IW	MW
Centerline Equipment						0	0	0
Layout foundations	0.12	3120.0	SF	0.0	SF			0
Set Lower Exhaust Hood								
Clean protective coating	0.25	104.0	LF	0.0	LF			0
Weld LO supply and drain downcomers	6.00	2.0	EA	0.0	EA			0
Bolt connections— mid standard	0.50	12.0	EA	0.0	EA			0
Fit mid-standard to sole plates	2.00	4.0	EA	0.0	EA			0
Install and grout retaining strips	0.50	104.0	LF	0.0	LF			0
Set lower turbine and exhaust hood	1.02	176.9	ton	0.0	ton		0	
Clean and set turning gear standard	80.00	1.0	EA	0.0	EA			0
Move exhaust hood	80.00	1.0	EA	0.0	EA			0
Rough alignment	1.42	176.9	ton	0.0	ton			0
Seal weld mid-standard	0.50	30.0	LF	0.0	LF		0	
Install Lower Inner Casing								
Rig and install inner casing	2.56	52.4	ton	0.0	ton			0
Install lower half packing casing	20.00	2.0	EA	0.0	EA			0
Rough alignment	1.42	52.4	ton	0.0	ton			0
Install Upper Inner Casing								
Rig and install upper casing	2.56	56.4	ton	0.0	ton			0
Bolt connection—inner casing	0.50	112.0	EA	0.0	EA			0
Final alignment	2.10	56.4	ton	0.0	ton			0
Crossover flange level check	8.00	1.0	EA	0.0	EA			0
Install Upper Half Exhaust Hood								
Install upper exhaust hood halves	2.56	77.0	ton	0.0	ton			0
Rough alignment	1.42	77.0	ton	0.0	ton			0
Bolt connection—vertical joint	0.50	60.0	EA	0.0	EA			0
Close Exhaust Hood								
Bolt connection—exhaust hood	0.50	216.0	EA	0.0	EA			0
Final alignment	2.10	176.9	ton	0.0	ton			0
Exhaust Hood Centerline Sole Plate								
Adjust left and right side anchor blocks	20.00	8.0	EA	0.0	EA			0
Weld anchor blocks exhaust hood	10.00	8.0	EA	0.0	EA			0
Tops-On Exhaust Hood Alignment								
Final alignment	2.10	77.0	ton	0.0	ton			0
Turning Gear Standard								
Final alignment	2.10	52.4	ton	0.0	ton			0
Weld Exhaust Hood to Condenser								
Weld exhaust hood condenser neck	1.50	104.0	LF	0.0	LF	0		
Alignment during welding	2.10	52.4	ton	0.0	ton	0		
Install condenser joint shield	40.00	1.0	EA	0.0	EA	0		
Final Grout Sole Plates								
Clean area and install grout sole plates	16.00	16.0	EA	0.0	EA			0
Final Tops-on Alignment								
Final alignment	2.10	52.4	ton	0.0	ton			0
Fit Exhaust Hood Keys								
Fit internal exhaust hood keys	1.00	40.0	EA	0.0	EA			0
Set and Fit Bearing Rings								
Set T2, T3, and T4 bearing rings	10.00	3.0	EA	0.0	EA			0

2.7.2 HP/IP Turbine and Front Standard Sheet 2

Estimate HP/IP Turbine and Front Standard	Historical			Estimate				
Description	MH	Quantity	Unit	Quantity	Unit	BM	IW	MW
HP/IP Turbine and Front Standard						0	0	0
Front Standard Base Plate Check								
Set fixators	3.50	4.0	EA	0.0	EA			0
Clean and set front standard plate	20.00	1.0	EA	0.0	EA			0
Front Standard Base Plate Check								
Rough alignment	1.42	32.0	ton	0.0	ton			0
Final alignment	2.10	32.0	ton	0.0	ton			0
Set HP/IP Turbine Assembly								
Set assembled HP/IP turbine	1.70	59.1	ton	0.0	ton		0	
Remove turbine end shipping brace	4.00	4.0	EA	0.0	EA		0	
Rough alignment	1.42	59.1	ton	0.0	ton			0

2.7.3 Final Turbine Assembly/Alignment Sheet 3

Estimate—Final Turbine Assembly		Historical		Estimate				
Description	MH	Quantity	Unit	Quantity	Unit	BM	IW	MW
Final Turbine Assembly and Alignment						0	0	0
Upper Exhaust Hood–Inner Casing								
Remove dowels from exhaust hood	0.55	216.0	EA	0.0	EA			0
Rig and remove upper exhaust hood	1.56	77.0	ton	0.0	ton			0
Remove Upper–Inner Casing								
Loosen and remove dowels and bolts	0.50	112.0	EA	0.0	EA			0
Rig and remove upper inner casing	2.56	56.4	ton	0.0	ton			0
Final Tops-Off Alignment								
Alignment (T3 and T4 inboard oil deflector)	2.10	13.8	ton	0.0	ton			0
Lower Half Low Pressure Diaphragms								
Clam protective coating	0.25	79.0	LF	0.0	LF			0
Rig and install diaphragms	14.00	12.0	EA	0.0	EA			0
Rough alignment	1.42	5.8	ton	0.0	ton			0
Number 3 and 4 Bearings								
Bearings, bearing rings, and standard fits	0.25	36.0	LF	0.0	LF			0
Install bearings	16.00	3.0	EA	0.0	EA			0
Install upper half bearing rings	16.00	3.0	EA	0.0	EA			0
Rough alignment	1.42	17.0	ton	0.0	ton			0
Low Pressure Rotor								
Clam protective coating	0.25	48.0	LF	0.0	LF			0
Rig and install low pressure rotor	1.65	84.9	ton	0.0	ton			0
Install Low Pressure Oil Deflectors								
Clam protective coating	0.25	36.0	LF	0.0	LF			0
Install lower half oil deflectors	10.00	12.0	EA	0.0	EA			0
Upper Half Pressure Bearings and Rings								
Install upper half bearings and rings	20.00	3.0	EA	0.0	EA			0
Rough alignment	0.04	600.0	lb	0.0	lb			0
Upper Half Low Pressure Diaphragms								
Clam protective coating	0.25	79.0	LF	0.0	LF			0
Rig and install diaphragms	14.00	12.0	EA	0.0	EA			0
Rough alignment	1.42	5.8	ton	0.0	ton			0
Upper Half Casing								
Rig and install the upper inner casing	2.56	56.4	ton	0.0	ton			0
Install horizontal joint shields	0.50	120.0	LF	0.0	LF			0
Final alignment	2.10	56.4	ton	0.0	ton			0
Final Turbine Assembly and Alignment								
Upper Half Exhaust Casing								
Rig and install the upper exhaust hood	1.56	77.0	ton	0.0	ton			0
Rough alignment	1.42	77.0	ton	0.0	ton			0
Bolt connection—vertical joint	0.50	60.0	EA	0.0	EA			0
Install atmosphere relief diaphragm	16.00	2.0	EA	0.0	EA			0
HP/IP–LP Assembly								
Set T2 lower half bearings	20.00	1.0	EA	0.0	EA			0
Rig and lift generator end of rotor	1.60	20.1	ton	0.0	ton			0
Remove shipping brackets	4.00	4.0	EA	0.0	EA			0
Install upper T1 and T2 bearings	20.00	2.0	EA	0.0	EA			0
Rough alignment	1.42	20.1	ton	0.0	ton			0

2.7.4 Turbine–Generator Acoustic Enclosure Sheet 4

Estimate—Generator Acoustic Enclosure		Historical			Estimate				
Description	MH	Quantity	Unit	Quantity	Unit	BM	IW	MW
Generator Acoustic Enclosure						0	0	0
Front Standard Base Plate								
Grout front standard base plate	1.40	43.0	SF	0.0	SF			0
Horn Drop Check								
Install/remove generator end rotor support	40.00	1.0	EA	0.0	EA			0
Final Alignment HP/IP Rotor								
Final alignment	2.10	105.0	ton	0.0	ton			0
"A" Coupling								
Clean coupling and associated hardware	0.25	32.0	LF	0.0	LF			0
Assemble, fit and dowel "A" coupling guard	30.00	1.0	EA	0.0	EA			0
Rough alignment	1.42	0.5	ton	0.0	ton			0
Install and fit "A" coupling covers	10.00	2.0	EA	0.0	EA			0
Install and fit upper half oil deflectors	10.00	2.0	EA	0.0	EA			0
Install HP/IP Centerline Keys								
Machine and install center keys	2.58	40.0	EA	0.0	EA			0
Rough alignment	1.42	1.0	ton	0.0	ton			0
Install crossover								
Remove shipping skid and coatings	0.25	32.0	LF	0.0	LF	0		
Remove expansion joint locking device	20.00	1.0	EA	0.0	EA	0		
54″ diameter fit up and weld-HP/LP turbine	1.15	54.0	DI	0.0	DI	0		
54″ diameter fir up and weld assembly	1.15	104.0	DI	0.0	DI	0		
Install r4″ diameter crossover pipe	3.64	32.0	LF	0.0	LF	0		
54″ diameter flange bolt up	0.45	54.0	DI	0.0	DI	0		
Install and align turning gear								
Install turning gear to rotor	40.00	1.0	EA	0.0	EA			0
Final alignment	2.10	13.7	ton	0.0	ton			0
Dowel turning gear to standard	1.20	16.5	EA	0.0	EA			0
Turbine Generator Sections								
Left lower quarter section	40.00	1.0	EA	0.0	EA		0	
Right lower quarter section	40.00	1.0	EA	0.0	EA		0	
Middle quarter section	40.00	2.0	EA	0.0	EA		0	
End half section	40.00	2.0	EA	0.0	EA		0	
Bottom plate	40.00	1.0	EA	0.0	EA		0	
Doors	8.00	4.0	EA	0.0	EA		0	
Field joints	16.00	12.0	EA	0.0	EA		0	

2.7.5 Generator Installation, Lube Oil System, Hydraulic Power Oil System Sheet 5

Estimate—Generator Installation		Historical		Estimate				
Description	MH	Quantity	Unit	Quantity	Unit	BM	IW	MW
Generator Installation						0	0	0
Generator terminal compartment								
Assemble cribbing and set compartment	120.0	1.0	EA	0.0	EA		0	
Rigging and moving generator								
Clean and install trunnions	10.0	8.0	EA	0.0	EA			0
Rough alignment stator								
Rig, lift, and move and set generator	1.4	259.5	ton	0.0	ton		0	
Remove and install generator outer shields	20.0	2.0	EA	0.0	EA		0	
Install generator journal bearings	10.0	4.0	EA	0.0	EA			0
Bolt-up outer end shield and dowel	1.0	48.0	EA	0.0	EA			0
Install hydrogen seal rings	25.0	1.0	EA	0.0	EA			0
Install and align oil deflectors	10.0	12.0	EA	0.0	EA			0
Set and rough alignment-státor	1.4	185.5	ton	0.0	ton			0
Final generator alignment								
Final alignment	2.1	259.5	ton	0.0	ton			0
Grout sole plate	16.0	8.0	EA	0.0	EA			0
LP turbine generator								
Install studs and align coupling guard	40.0	1.0	EA	0.0	EA			0

Estimate—LO System		Historical		Estimate				
Lube Oil System						0	0	0
Set lube oil tank	60.0	1.0	EA	0.0	EA	0		
Align lube oil pumps	20.0	3.0	EA	0.0	EA	0		
Install lube oil piping	2.2	100.0	LF	0.0	LF	0		
Install lube oil conditioner	60.0	1.0	EA	0.0	EA	0		

Estimate—Hydraulic Power Oil System		Historical		Estimate				
Hydraulic Power Oil System						0	0	0
Set hydraulic power unit	80.0	1.0	EA	0.0	EA	0		
Align hydraulic power pumps	20.0	3.0	EA	0.0	EA	0		
Install hydraulic power piping	2.2	200.0	LF	0.0	LF	0		

2.8   Reheat Double Flow STG-Installation Man Hours

Facility-Combined Cycle Power Plant	Actual	Estimated			
Description	MH	BM	IW	MW	MH
Centerline equipment	3729	0	0	0	**0**
HP/IP turbine and front standard	347	0	0	0	**0**
Final turbine assembly and alignment	2106	0	0	0	**0**
Generator acoustic enclosure	1447	0	0	0	**0**
Generator installation	1811	0	0	0	**0**
Lube oil system	400	0	0	0	**0**
Hydraulic power oil system	580	0	0	0	**0**
Equipment installation man hours	**10420**	**0**	**0**	**0**	**0**

2.9   General Scope of Field Work Required for Each Three-Wide HRSG

HRSG—Triple Pressure w/Reheat for F-Class GT
Scope of Work-Field Erection

Erection of HRSG casing
Install foundation slide plates
Erect primary casing panels
    Install prefabricated casing floor panels
Erect secondary casing panels
Install field seams along column lines
    Install floor and sidewall field seams
Moment welds
Install CO catalyst internal support structure
Erection of HRSG ducting
Erect inlet duct, SCR duct, and outlet transition duct
    Assemble and set inlet ducts
    Install inlet duct floor, sidewall, and roof casing seams
Moment welds
Install SCR catalyst internal support structure with the SCR duct
    SCR reactor housing and SCR catalyst support structure
    Platforms, walkways, and ladders supplied by shop fabricator
    Install stair tower, ladders, and platforms
Installation of modules
    All modules shipped with roof casing shop installed
    All roof casing penetrations/packing glands shop installed
    Unload modules
    Set up lifting sled for each module
    Lift and install modules
Installation of gas baffles
Install gas baffles in front and back of each module
Install HP, IP, and LP steam drums
Install skids
HRSG—Install duct burner assemblies and elements
Duct burner assembly
Erect stack, breeching, and outlet transition duct
Erect stack
Erect breeching and outlet transition duct
    Assemble and install outlet breeching duct panels
    Assemble and install exhaust stack
    Install outlet expansion joints

2.10   HRSG-Triple Pressure, Three-Wide Estimate Data

2.10.1   Module Casing Sheet 1

Description	MH	Unit
Slide Plate Assemblies		
Shim and set slide/base plates (provide shims)	1.00	EA
Weld shear blocks	0.50	EA
Grout base plates	4.00	EA
Anchor bolts	0.65	EA
Erect Casing Panels and Assemblies		
Casing assembly columns, floor beams	20.00	EA
Roof beam corner angle	4.00	EA
Bellows	2.60	EA
Roof beams assembly	25.00	EA
Casing assembly; flange assembly; grid Assembly	10.00	EA
Floor panel	10.00	EA
Moment Welds		
Moment weld-column 13/16″ single groove/3/16″ fillet	12.00	EA
Moment weld-column 9/16″ single groove/3/16″ fillet	10.00	EA
Moment weld-column 5/8″ single groove/3/16″ fillet	8.00	EA
Bolt up 7/8′′ diameter×2″	1.00	EA
Fillet weld at web 1/4″	2.00	EA
Remove lift lug	0.50	EA
Remove angle clip	0.50	EA
Field Seam Liners and Insulation		
Install ceramic fiber blanket	0.04	SF
Install ogee clips, washers, bolt and weld washer	0.25	EA
Install mesh SS screen wire	0.04	SF
Install insulation and liner plates	1.00	SF
Weld Casing Seams		
Seal weld side, roof and floor casing field seams	0.35	LF
Fit-up	0.05	LF

2.10.2   Duct Work and SCR/CO Sheet 2

Description	MH	Unit
Slide Plate Assemblies		
Shim and set slide/base plates (provide shims)	1.00	EA
Weld shear blocks	0.50	EA
Grout base plates	4.00	EA
Anchor bolts	0.65	EA
Erect Casing Panels and Assemblies		
Casing assembly columns, floor beams	20.00	EA
Roof beam corner angle	4.00	EA
Bellows	2.60	EA
Roof beams assembly	25.00	EA
Casing assembly; flange assembly; grid assembly	10.00	EA
Floor panel	10.00	EA
Moment Welds		
Moment weld-column 13/16″ single groove/3/16″ fillet	12.00	EA
Moment weld-column 9/16″ single groove/3/16″ fillet	10.00	EA
Moment weld-column 5/8″ single groove/3/16″ fillet	8.00	EA
Bolt up 7/8′′ diameter×2″	1.00	EA
Fillet weld at web 1/4″	2.00	EA
Remove lift lug	0.50	EA
Remove angle clip	0.50	EA
Field Seam Liners and Insulation		
Install ceramic fiber blanket	0.04	SF
Install ogee clips, washers, bolt and weld washer	0.25	EA
Install mesh SS screen wire	0.04	SF
Install insulation and liner plates	1.00	SF
Weld Casing Seams		
Seal weld side, roof and floor casing field seams	0.35	LF
Fit-up	0.05	LF

2.10.3 Baffles, Modules, Pressure Vessels and Duct Burner Sheet 3

Description	MH	Unit
Erect Gas Baffles		
Baffle plates	0.75	EA
Fillet weld	0.35	LF
Splice plates	0.75	EA
U-bolt connections	0.35	EA
Header/shipping restraints	1.00	EA
Modules		
75,000 lb	2.5	ton
75,001 lb > weigh t < 125,000 lb	2.2	ton
125,001 lb > weight <= 200,000 lb	1.8	ton
200,001 lb > weight <= 300,000 lb	1.5	ton
300,001 lb > weight <= 400,000 lb	1.4	ton
Pressure Vessels		
75,000 lb	2.5	ton
75,001 lb > weight < 125,000 lb	2.2	ton
125,001 lb > weight <= 200,000 lb	1.8	ton
200,001 lb > weight <= 300,000 lb	1.6	ton
300,001 lb > weight <= 400,000 lb	1.4	ton
Set saddle base plates	24.00	EA
Install Duct burner Assemblies and Elements		
Duct burner assembly, headers and element assembly	20.00	EA
Scanner, hose, and retainer assembly	10.00	EA
Element support	5.00	EA

2.10.4 Platforms, Skids, and SCR/CO Internals Sheet 4

Description	MH	Unit
Steel (Ladders, Platforms, and Grating)		
Assemble and install structural steel and platforms	0.40	SF
Handrail LF	0.32	LF
Ladders LF	1.00	LF
Stair tower structural	32.00	ton
Skids		
Level and set skids	80.00	EA
SCR and CO$_2$ Internals		
Set support structure and cut/weld and grind lifting lugs	30.00	EA
Set left, middle, and right internal structure	30.00	EA
Weld structure bases to each other	0.35	LF
Insert and weld top and side pins-4″–1/4″ fillet weld	0.35	EA
Catalyst Anchor Installation		
Set anchor channel and weld to structure	0.35	LF
Install T-shaped anchors	0.75	EA
Make-up T-shaped-washer/nut w/jam nut	0.35	EA
Seal Plate Installation		
Install side/top/bottom slide plates	2.00	EA
Weld seal plates	0.35	LF
Trim seal plates	0.25	LF
Remove shipping braces	1.00	EA

2.10.5   Stack and Breeching Sheet 5

Description	MH	Unit
Stack and Breeching (XX'-X'' OD × XXX'-X'' High)		
Set shell cans (field fabrication)	4.00	EA
Weld vertical (shell can to shell can w/stiffener splices)	0.75	LF
Weld horizontal	0.50	LF
Shoot Shim Packs, Erect Lower/Upper Stack, and Grout Base Ring		
Shims and bolt to foundation	2.40	EA
Set lower stack section	60.00	EA
Load and haul to erection site	4.00	EA
Erect Damper		
Seam up (2) damper halves		
Set up damper halves (field fabrication)	10.00	EA
Weld damper halves	0.80	LF
Bolt up	0.40	EA
Attach Structural Steel and Platforms		
360 degree platform	0.15	SF
Ladder	0.35	LF
360 degree platform	0.15	SF
Bridge	0.15	SF
Fit and Weld Breech Sections to Stack		
Set panels	60.00	EA
Weld	0.75	LF
Install Expansion Joint		
Expansion joint	60.00	EA
Backup bars	1.00	EA
Bolt up	0.15	EA
Expanded metal	1.20	EA
Load and haul dampers to erection site	16.00	EA

2.11 HRSG-Triple Pressure, Three-Wide Estimate

2.11.1 Module Casing Sheet 1

Estimate—Module Casing	Historical			Estimate		
Description	MH	Quantity	Unit	Quantity	Unit	BM
HRSG—Module Casing						**0**
Slide Plate Assemblies						
Shim and set slide/base plates (provide shims)	1.00	45.0	EA	0.0	EA	0
Weld shear blocks	0.50	180.0	EA	0.0	EA	0
Grout base plates	4.00	45.0	EA	0.0	EA	0
Anchor bolts	0.65	180.0	EA	0.0	EA	0
Erect Casing Panels and Assemblies						
Casing assembly columns, floor beams	20.00	35.0	EA	0.0	EA	0
Roof beam corner angle	4.00	24.0	EA	0.0	EA	0
Bellows	2.60	40.0	EA	0.0	EA	0
Install Roof/Floor Beam Assemblies						
Roof Beams Assembly	25.00	24.0	EA	0.0	EA	0
Frame Connections						
Moment Welds						
Moment weld-col 13/16″ single groove/3/16″ fillet	12.00	128.0	EA	0.0	EA	0
Bolt up 7/8′ ′ diameter×2″	1.00	96.0	EA	0.0	EA	0
Fillet weld at web 1/4″	2.00	64.0	EA	0.0	EA	0
Remove lift lug	0.50	32.0	EA	0.0	EA	0
Remove angle clip	0.50	32.0	EA	0.0	EA	0
Field Seam Liners and Insulation						
Install ceramic fiber blanket	0.04	4663.4	SF	0.0	SF	0
Install ogee clips, washers, bolt and weld washer	0.25	460.0	EA	0.0	EA	0
Install mesh SS screen wire	0.04	4663.4	SF	0.0	SF	0
Install insulation and liner plates	1.00	460.0	SF	0.0	SF	0
Weld Casing Seams						
Casing assembly w/cols	0.35	1537.0	LF	0.0	LF	0
Casing assembly w/floor beams	0.35	897.0	LF	0.0	LF	0
Roof casing w/tube bundle assembly	0.35	1255.8	LF	0.0	LF	0
Casing panel	0.35	973.6	LF	0.0	LF	0
Fit-up	0.05	4663.4	LF	0.0	LF	0

2.11.2   Duct Work SCR/CO Sheet 2

Estimate—Duct Work and SCR/CO		Historical		Estimate		
Description	MH	Quantity	Unit	Quantity	Unit	BM
HRSG—Duct Work SCR and Inlet						**0**
Slide Plate Assemblies (SCR and Inlet)						
Shim and set slide/base plates (provide shims)	1.00	12.0	EA	0.0	EA	0
Weld sheer blocks	0.50	48.0	EA	0.0	EA	0
Grout base plates	4.00	12.0	EA	0.0	EA	0
Anchor bolts	0.65	48.0	EA	0.0	EA	0
Erect Casing Panels and Assemblies						
Casing assembly, flange assembly, grid assembly	10.00	24.0	EA	0.0	EA	0
Floor panel	10.00	8.0	EA	0.0	EA	0
Casing assembly w/roof beam	20.00	6.0	EA	0.0	EA	0
Install Roof/Floor Beam Assemblies						
SCR						
Roof beams assembly	25.00	6.0	EA	0.0	EA	0
Moment Welds						
Frame connection: inlet duct "A," "B," "C," and "D"						
Moment weld-col 13/16″″ single groove/3/16″ fillet	12.00	48.0	EA	0.0	EA	0
Moment weld-col 9/16″″″ single groove/3/16″ fillet	10.00	32.0	EA	0.0	EA	0
Moment weld-col 5/8″″″ single groove/3/16″ fillet	8.00	16.0	EA	0.0	EA	0
Bolt up 7/8″′ ″″ diameter×2″	1.00	48.0	EA	0.0	EA	0
Fillet weld at web 1/4″″	2.00	31.5	EA	0.0	EA	0
Remove lift lug	0.50	32.0	EA	0.0	EA	0
Remove angle clip	0.50	48.0	EA	0.0	EA	0
Field Seam Liners and Insulation						
Inlet Duct						
Field Joint at Wall						
Install insulation and liner plates	1.00	150.0	EA	0.0	EA	0
Install ogee clips, washers, bolt and weld washer	0.25	150.0	EA	0.0	EA	0
Field Joint at Floor and Roof						
Install insulation and liner plates	1.00	120.0	EA	0.0	EA	0
Install ogee clips, washers, bolt and weld washer	0.25	120.0	EA	0.0	EA	0

Continued

Estimate—Duct Work and SCR/CO	Historical			Estimate		
Description	MH	Quantity	Unit	Quantity	Unit	BM
SCR/CO						
Insulate and Install Lower Liner Plates at Field Seams						
Plates	1.00	400.0	EA	0.0	EA	0
Insulation	0.04	4800.0	SF	0.0	SF	0
Weld Casing Seams						
Spool Duct						
Casing panel	0.35	307.4	LF	0.0	LF	0
Casing panel	0.35	175.4	LF	0.0	LF	0
Roof casing w/tube bundle assembly	0.35	175.4	LF	0.0	LF	0
SCR						
Casing assembly w/cols	0.35	307.4	LF	0.0	LF	0
Casing assembly w/floor beams	0.35	165.4	LF	0.0	LF	0
Roof casing w/tube bundle assembly	0.35	165.4	LF	0.0	LF	0
Inlet Duct						
Field Joint at Wall						
Seal weld secondary casing to column	0.35	440.0	LF	0.0	LF	0
Field Joint at Floor and Roof						
Seal weld secondary casing to column	0.35	270.0	LF	0.0	LF	0
Cover plate-floor/roof	0.35	270.0	LF	0.0	LF	0

2.11.3 Baffles, Modules, Pressure Vessels, and Duct Burner Sheet 3

Estimate—Baffles	Historical			Estimate		
Description	MH	Quantity	Unit	Quantity	Unit	BM
HRSG—Erect Gas Baffles						**0**
Install Side Wall Gas Baffles						
Baffle plates	0.75	880.0	EA	0.0	EA	0
Fillet weld	0.35	1320.0	LF	0.0	LF	0
Install Upper Splice Gas Baffles						
Splice plates	0.75	647.0	EA	0.0	EA	0
U-bolt connections	0.35	647.0	EA	0.0	EA	0
Install Lower Gas Baffles						
Baffle plates	0.75	307.0	EA	0.0	EA	0
U-bolt connections	0.35	307.0	EA	0.0	EA	0
Header/Shipping Restraints						
Angle 4×4×3/4×6″-1-5/8″″	1.00	78.0	EA	0.0	EA	0
Tube Bundle Gas Baffles						
Baffle plates	0.75	299.0	EA	0.0	EA	0
Fillet weld	0.35	299.0	LF	0.0	LF	0

Estimate—Modules	Historical			Estimate		
HRSG—Modules						**0**
Reheater #3 hp Superheated #2/Reheater #2	2.20	56.0	ton	0.0	ton	0
HP Superheated #1/Reheater #1	1.80	65.5	ton	0.0	ton	0
HP evaporator/IP superheated #2	1.40	165.5	ton	0.0	ton	0
HP economizer #2/IP superheated #1	1.50	147.0	ton	0.0	ton	0
HP economizer #1/IP economizer #1/LP evaporator	1.50	138.5	ton	0.0	ton	0
Feed water heater #2	2.20	55.0	ton	0.0	ton	0
Feed water heater #1	2.50	35.5	ton	0.0	ton	0

Estimate—Pressure Vessels	Historical			Estimate		
HRSG—Pressure Vessels						**0**
HP Steam Drum						
Set saddle base plates	24.00	1.0	EA	0.0	EA	0
Set drum	1.40	158.0	ton	0.0	ton	0
IP Steam Drum						
Set saddle base plates	24.00	1.0	EA	0.0	EA	0
Set drum	2.50	9.0	ton	0.0	ton	0
LP Steam Drum						
Set saddle base plates	24.00	1.0	EA	0.0	EA	0
Set drum	2.20	30.0	ton	0.0	ton	0
Flash Separator						
Set saddle base plates	24.00	1.0	EA	0.0	EA	0
Set drum	2.50	6.0	ton	0.0	ton	0

Estimate—Duct Burner	Historical			Estimate		
HRSG—Install Duct Burner Assemblies and Elements						**0**
Duct burner assembly	20.00	18.0	EA	0.0	EA	0
Scanner, hose, and retainer assembly	10.00	55.0	EA	0.0	EA	0
Element support	5.00	50.0	EA	0.0	EA	0

2.11.4 Platforms, Skids, and SCR/CO Internals Sheet 4

Estimate—Platforms		Historical		Estimate		
Description	MH	Quantity	Unit	Quantity	Unit	BM
HRSG—Steel (Ladders, Platforms, and Grating)						
Assemble and Install all of the						**0**
Structural Steel and Platforms						
Top of HRSG L″×W′	0.40	4140.0	SF	0.0	SF	0
HP, IP, LP drums L′×W′	0.40	540.0	SF	0.0	SF	0
Top of inlet duct L′×W′	0.40	1080.0	SF	0.0	SF	0
Right/left side X EA times (L′×W′)	0.40	518.4	SF	0.0	SF	0
Stair tower X EA times (L′×W′)	0.40	384.0	SF	0.0	SF	0
Misc X EA times (L′×W′)	0.40	216.0	SF	0.0	SF	0
Handrail LF	0.32	714.0	LF	0.0	LF	0
Ladders LF	1.00	360.0	LF	0.0	LF	0
Stair tower structural	32.00	30.0	ton	0.0	ton	0

Estimate—Skids		Historical		Estimate		
HRSG—Skids						**0**
Level and set the AFCU skid	80.00	1.0	EA	0.0	EA	0
Level and set the piping module	80.00	1.0	EA	0.0	EA	0
Level and set the blower skid	80.00	1.0	EA	0.0	EA	0

Estimate—SCR and CO		Historical		Estimate		
HRSG—SCR and CO$_2$ Internals						**0**
Install Internal Structure						
Remove shipping braces	1.00	100.0	EA	0.0	EA	0
Set support structure and cut/weld and grind lifting lugs	30.00	2.0	EA	0.0	EA	0
Set left, middle right internal structure	30.00	6.0	EA	0.0	EA	0
Weld structure bases to each other	0.35	560.0	LF	0.0	LF	0
Insert and weld top and side pins-4′-1/4″ fillet weld	0.35	18.0	EA	0.0	EA	0
Catalyst Anchor Installation						
Set anchor channel and weld to structure	0.35	216.0	LF	0.0	LF	0
Install T-shaped anchors	0.75	216.0	EA	0.0	EA	0
Make-up T-shaped-washer/nut w/jam nut	0.35	216.0	EA	0.0	EA	0
Seal Plate Installation						
Install side/top/bottom slide plates	2.00	62.0	EA	0.0	EA	0
Weld seal plates	0.35	1060.0	LF	0.0	LF	0
Trim seal plates	0.25	265.0	LF	0.0	LF	0

Estimate—Stack and Breeching		Historical		Estimate		
Description	MH	Quantity	Unit	Quantity	Unit	BM
HRSG—Stack and Breeching (XX′-X″ OD × XXX′-X″ High)						0
Erect lower stack						
Seam up shell cans for lower section						
Set shell cans (field fabrication)	4.00	16.0	EA	0.0	EA	0
Weld vertical (shell can to shell can w/stiffener splices)	0.75	224.0	LF	0.0	LF	0
Stack shell cans and weld to make-up lower sections						
Set shell cans	4.00	16.0	EA	0.0	EA	0
Weld horizontal	0.75	450.0	LF	0.0	LF	0
Shoot Shim Packs, Erect Lower Stack, and Grout Base Ring						
Shims and bolt to foundation	2.40	32.0	EA	0.0	EA	0
Set lower stack section	60.00	1.0	EA	0.0	EA	0
Load and haul to erection site	4.00	28.0	EA	0.0	EA	0
Erect Damper						
Seam up (2) damper halves						
Set up damper halves (field fabrication)	10.00	2.0	EA	0.0	EA	0
Weld damper halves	0.80	20.0	LF	0.0	LF	0
Bolt up	0.40	100.0	EA	0.0	EA	0
Load and haul dampers to erection site	16.00	2.0	EA	0.0	EA	0
Erect Upper Stack						
Seam Up Shell Cans for Upper Section						
Set shell cans (field fabrication)	4.00	14.0	EA	0.0	EA	0
Weld vertical (shell can to shell can w/stiffener splices)	0.75	220.0	LF	0.0	LF	0
Stack Shell Cans and Weld to Make-Up Upper Sections						
Set shell cans	4.00	16.0	EA	0.0	EA	0
Weld horizontal	0.50	460.0	LF	0.0	LF	0
Erect Upper Stack Sections						
Set upper sections	60.00	1.0	EA	0.0	EA	0
Weld horizontal	0.75	120.0	LF	0.0	LF	0
Load and haul to erection site	4.00	20.0	EA	0.0	EA	0
Attach Structural Steel and Platforms						
360 degree platform	0.15	360.0	EA	0.0	EA	0
Ladder	0.35	132.0	LF	0.0	LF	0
360 degree platform	0.15	360.0	SF	0.0	SF	0
Bridge	0.15	240.0	SF	0.0	SF	0
Fit and Weld Breech Sections to Stack						
Set panels	60.00	2.0	EA	0.0	EA	0
Weld	0.75	330.0	LF	0.0	LF	0
Install Expansion Joint						
Expansion joint	60.00	1.0	EA	0.0	EA	0
Backup bars	1.00	24.0	EA	0.0	EA	0
Bolt up	0.25	1120.0	EA	0.0	EA	0
Expanded metal	1.20	200.0	EA	0.0	EA	0

2.12 HRSG Triple Pressure; Three-Wide-Installation Man Hours

Facility-Combined Cycle Power Plant	Actual	Estimated
Description	MH	Unit
HRSG—module casing	6528	0
HRSG—duct work SCR and inlet	3606	0
HRSG—erect gas baffles	2578	0
HRSG—modules	1111	0
HRSG—pressure vessels	420	0
HRSG—install duct burner assemblies and elements	1160	0
HRSG—steel (ladders, platforms, and grating)	4300	0
HRSG—skids	240	0
HRSG—stack and breeching	2898	0
HRSG—SCR and CO_2 internals	1417	0
Equipment installation man hours	**24258**	**0**

2.13 General Scope of Field Work Required for Each Double Wide HRSG

HRSG—Triple Pressure w/Reheat for F-Class GT
Scope of Work-Field Erection

Erection of HRSG Casing
Install foundation slide plates
Erect primary casing panels
    Install prefabricated casing floor panels
Erect secondary casing panels
Install field seams along column lines
    Install floor and sidewall field seams
Moment welds
Install CO catalyst internal support structure
Erection of HRSG Ducting
Erect inlet duct, SCR duct, and outlet transition duct
    Assemble and set inlet ducts
    Install inlet duct floor, sidewall, and roof casing seams
Moment welds
Install SCR catalyst internal support structure with the SCR duct
    SCR reactor housing and SCR catalyst support structure
    Platforms, walkways, and ladders supplied by shop fabricator
    Install stair tower, ladders, and platforms
Installation of Modules
    All modules shipped with roof casing shop installed
    All roof casing penetrations/packing glands shop installed
    Unload modules
    Set up lifting sled for each module
    Lift and install modules
Installation of gas baffles
Install gas baffles in front and back of each module
Install HP, IP, and LP steam drums
Install skids
HRSG—install duct burner assemblies and elements
Duct burner assembly
Erect Stack, breeching, and outlet transition duct
Erect stack
Erect breeching and outlet transition duct
    Assemble and install outlet breeching duct panels
    Assemble and install exhaust stack
    Install outlet expansion joints

2.14 HRSG Triple Pressure Double Wide Estimate Data

2.14.1 Module Casing Sheet 1

Description	MH	Unit
Slide Plate Assemblies		
Shim and set slide/base plates (provide shims)	1.00	EA
Weld shear blocks	0.50	EA
Grout base plates	4.00	EA
Anchor bolts	0.65	EA
Erect Casing Panels and Assemblies		
Casing assemblies and floor panels	20.00	EA
Load and haul to erection site	8.00	EA
Roof beams assembly	25.00	EA
Casing assembly	10.00	EA
Floor panel	10.00	EA
Erect Roof Filler Panels		
Seal weld side, roof, wall, corner, and cover casing plate	0.54	LF
Frame Connections		
Roof beams assembly		
WT	4.0	EA
Hex bolts 1 1/4″ diameter	1.0	EA
Floor Beam		
WT	4.0	EA
Hex bolts 1 1/4″ diameter	1.0	EA
5/8″ single groove weld/1/2″ fillet	8.0	EA
Field Seam Liners and Insulation		
Install ceramic fiber blanket	0.04	SF
Install ogee clips, washers, bolt and weld washer	0.25	EA
Install mesh SS screen wire	0.04	SF
Install insulation and liner plates	1.00	SF
Weld Casing Seams		
Seal weld side, roof and floor casing field seams	0.35	LF
Fit-up	0.05	LF
Moment Welds		
Frame Connection: Inlet Duct "A," "B," and "C"		
Moment weld-col; 1/2″/3/16″ single groove/3/16″ fillet	8.00	LF
Moment weld-col; 3/4″/3/8″ single groove/3/16″ fillet	10.00	LF
Moment weld-col; 2″/1–15/16″ single groove/3/16″ fillet	16.00	LF
Bolt up-col 3; 7/8″ diameter × 2	1.00	EA
Fillet weld at web 7/16″	2.00	EA
Remove lift lug	0.50	EA

2.14.2   Duct Work SCR and Inlet Sheet 2

Description	MH	Unit
Slide Plate Assemblies		
Shim and set slide/base plates (provide shims)	1.00	EA
Weld shear blocks	0.50	EA
Grout base plates	4.00	EA
Anchor bolts	0.65	EA
Erect Casing Panels and Assemblies		
Casing assemblies and floor panels	20.00	EA
Load and haul to erection site	8.00	EA
Roof beams assembly	25.00	EA
Casing assembly	10.00	EA
Floor panel	10.00	EA
Erect Roof Filler Panels		
Seal weld side, roof, wall, corner, and cover casing plate	0.54	LF
Frame Connections		
Roof beams assembly		
WT	4.0	EA
Hex bolts 1 1/4″ diameter	1.0	EA
Floor Beam		
WT	4.0	EA
Hex bolts 1 1/4″ diameter	1.0	EA
5/8″ single groove weld/1/2″ fillet	8.0	EA
Field Seam Liners and Insulation		
Install ceramic fiber blanket	0.04	SF
Install ogee clips, washers, bolt and weld washer	0.25	EA
Install mesh SS screen wire	0.04	SF
Install insulation and liner plates	1.00	SF
Weld Casing Seams		
Seal weld side, roof, and floor casing field seams	0.35	LF
Fit-up	0.05	LF
Moment Welds		
Frame Connection: Inlet Duct "A," "B," and "C"		
Moment weld-col; 1/2″/3/16″ single groove/3/16″ fillet	8.00	LF
Moment weld-col; 3/4″/3/8″ single groove/3/16″ fillet	10.00	LF
Moment weld-col; 2″/1–15/16″ single groove/3/16″ fillet	16.00	LF
Bolt up -col 3; 7/8″ diameter × 2	1.00	EA
Fillet weld at web 7/16″	2.00	EA
Remove lift lug	0.50	EA

2.14.3   Baffles, Modules, Pressure Vessel Sheet 3

Description	MH	Unit
Erect Gas Baffles		
Baffle plates	0.75	EA
Fillet weld	0.35	LF
Splice plates	0.75	EA
U-bolt connections	0.35	EA
Header/shipping restraints	1.00	EA
Modules		
75,000 lb	2.5	ton
75,001 lb > weight < 125,000 lb	2.2	ton
125,001 lb > weight <= 200,000 lb	1.8	ton
200,001 lb > weight <= 300,000 lb	1.5	ton
300,001 lb > weight <= 400,000 lb	1.4	ton
Pressure Vessels		
75,000 lb	2.5	ton
75,001 lb > weight < 125,000 lb	2.2	ton
125,001 lb > weight <= 200,000 lb	1.8	ton
200,001 lb > weight <= 300,000 lb	1.5	ton
300,001 lb > weight <= 400,000 lb	1.4	ton
Set saddle base plates	24.00	EA

2.14.4   Platforms, Skids and SCR/CO Internals Sheet 4

Description	MH	Unit
Steel (Ladders, Platforms, and Grating)		
Assemble and install structural steel and platforms	0.40	SF
Handrail LF	0.32	LF
Ladders LF	1.00	LF
Stair tower structural	32.00	ton
Skids		
Level and set skids	80.00	EA
SCR and CO_2 Internals		
Set support structure and cut/weld and grind lifting lugs	30.00	EA
Set left, middle, and right internal structure	30.00	EA
Weld structure bases to each other	0.35	LF
Insert and weld top and side pins-4″-1/4″ fillet weld	0.35	EA
Catalyst Anchor Installation		
Set anchor channel and weld to structure	0.35	LF
Install T-shaped anchors	0.75	EA
Make-up T-shaped-washer/nut w/jam nut	0.35	EA
Seal Plate Installation		
Install side/top/bottom slide plates	2.00	EA
Weld seal plates	0.35	LF
Trim seal plates	0.25	LF
Remove shipping braces	1.00	EA

2.14.5   Stack and Breeching Sheet 5

Description	MH	Unit
Stack and Breeching (XX″-X ″ OD × XXX″-X″ High)		
Set shell cans (field fabrication)	4.00	EA
Weld vertical (shell can to shell can w/stiffener splices)	0.75	LF
Weld horizontal	0.50	LF
Shoot Shim Packs, Erect Lower/Upper Stack and Grout Base Ring		
Shims and bolt to foundation	1.80	EA
Set lower stack section	60.00	EA
Load and haul to erection site	2.00	ton
Erect Damper		
Seam up (2) damper halves		
Set up damper halves (field fabrication)	10.00	EA
Weld damper halves	0.80	LF
Bolt up	0.40	EA
Attach Structural Steel and Platforms		
360 degree platform	0.15	SF
Ladder	0.35	LF
360 degree platform	0.15	SF
Bridge	0.15	SF
Fit and Weld Breech Sections to Stack		
Set panels	60.00	EA
Weld	0.75	LF
Install Expansion Joint		
Expansion joint	60.00	EA
Backup bars	1.00	EA
Bolt up	0.15	EA
Expanded Metal	1.20	EA

2.15 HRSG-Triple Pressure, Double Wide Estimate

2.15.1 Module Casing Sheet 1

Estimate—Module Casing		Historical			Estimate		
Description	MH	Quantity	Unit	Quantity	Unit	BM
HRSG—Module Casing						**0**
Slide Plate Assemblies						
Shim and set slide/base plates (provide shims)	1.00	30.0	EA	0.0	EA	0
Weld shear blocks	0.50	120.0	EA	0.0	EA	0
Grout base plates	4.00	30.0	EA	0.0	EA	0
Anchor bolts	0.65	120.0	EA	0.0	EA	0
Erect Casing Panels and Assemblies						
Casing assembly w/cols	20.00	8.0	EA	0.0	EA	0
Casing assembly w/floor beams	20.00	8.0	EA	0.0	EA	0
Casing panel	20.00	8.0	EA	0.0	EA	0
Load and haul to erection site	8.00	34.0	EA	0.0	EA	0
Install Roof/Floor Beam Assemblies						
Roof beams assembly	25.00	9.0	EA	0.0	EA	0
Erect Roof Filler Panels						
Roof to wall corner-seal weld; 3/16″ fillet weld	0.54	132.0	LF	0.0	LF	0
Corner angle; bolt up-seal weld	0.54	132.0	LF	0.0	LF	0
Cover plate-roof; field cut/3/16″ fillet weld	0.54	48.0	LF	0.0	LF	0
Casing cover plate-roof; field cut/3/16″ fillet weld	0.54	132.0	LF	0.0	LF	0
Frame Connections						
Moment Welds						
Roof beams assembly						
WT 6×60×3′-4 1/4″	4.00	20.0	EA	0.0	EA	0
Hex bolts 1 1/4″ diameter	1.00	400.0	EA	0.0	EA	0
Floor Beam						
WT 6×76×2′-7″	4.00	20.0	EA	0.0	EA	0
Hex bolts 1 1/4″ diameter	1.00	400.0	EA	0.0	EA	0
5/8″ single groove weld/1/2″ fillet	8.00	20.0	EA	0.0	EA	0
Field Seam Liners and Insulation						
Install ceramic fiber blanket	0.04	8250.0	SF	0.0	SF	0
Install ogee clips, washers, bolt and weld washer	0.25	930.0	EA	0.0	EA	0
Install mesh SS screen wire	0.04	760.0	SF	0.0	SF	0
Install insulation and liner plates	1.00	566.0	EA	0.0	EA	0
Weld Casing Seams						
Casing assembly w/cols	0.35	1056.0	LF	0.0	LF	0
Casing assembly w/floor beams	0.35	354.0	LF	0.0	LF	0
Roof casing w/tube bundle assembly	0.35	530.0	LF	0.0	LF	0
Casing panel	0.35	705.0	LF	0.0	LF	0
Fit-up	0.05	2645.0	LF	0.0	LF	0

2.15.2 Duct Work SCR and Inlet Sheet 2

Estimate—Duct Work and SCR/CO	Historical			Estimate		
Description	MH	Quantity	Unit	Quantity	Unit	BM
HRSG—Duct Work SCR and Inlet						**0**
Slide Plate Assemblies (SCR and Inlet)						
Shim and set slide/base plates (Provide shims)	1.00	12.0	EA	0.0	EA	0
Weld sheer blocks	0.50	48.0	EA	0.0	EA	0
Grout base plates	4.00	12.0	EA	0.0	EA	0
Anchor bolts	0.65	48.0	EA	0.0	EA	0
Erect Casing Panels and Assemblies						
Casing assembly W/column/casing panel	20.00	10.0	EA	0.0	EA	0
Casing assembly w/floor beams/floor panel	20.00	5.0	EA	0.0	EA	0
Casing assembly w/roof beam	20.00	8.0	EA	0.0	EA	0
Install Roof/Floor Beam Assemblies						
Corner-seal weld; 3/16″ Fillet weld	0.54	242.0	LF	0.0	LF	0
Corner angle; bolt up-seal weld	0.54	16.0	LF	0.0	LF	0
Moment Welds						
Frame Connection: Inlet Duct "A," "B," and "C"						
Moment weld-column; 1/2″/3/16″ groove/3/16″ fillet	8.00	8.0	LF	0.0	LF	0
Moment weld-column; 3/4″/3/8″ groove/3/16″ fillet	10.00	8.0	LF	0.0	LF	0
Moment weld-column; 2‴/1–15/16″ groove/3/16″ fillet	16.00	8.0	LF	0.0	LF	0
Bolt up-column 3; 7/8″ diameter×2	1.00	12.0	EA	0.0	EA	0
Fillet weld at web 7/16″	2.00	12.0	EA	0.0	EA	0
Remove lift lug	0.50	6.0	EA	0.0	EA	0
Remove angle clip	0.50	6.0	EA	0.0	EA	0
Field Seam Liners and Insulation						
Inlet Duct						
Field Joint at Wall						
Install insulation and liner plates	1.00	48.0	EA	0.0	EA	0
Install ogee clips, washers, bolt and weld washer	0.25	1040.0	EA	0.0	EA	0
Field Joint at Floor and Roof						
Install insulation and liner plates	1.00	28.0	EA	0.0	EA	0
Install ogee clips, washers, bolt and weld washer	0.25	140.0	EA	0.0	EA	0
SCR/CO						
Insulate and Install Lower Liner Plates at Field Seams						
Plates	1.00	200.0	EA	0.0	EA	0
Insulation	0.04	2400.0	SF	0.0	SF	0
Weld Casing Seams						
Field Joint at Wall; Fillet Weld						
Seal weld secondary casing to column	0.35	332.0	LF	0.0	LF	0
Seal weld secondary casing to column-fillet weld	0.35	180.0	LF	0.0	LF	0
Cover plate-floor/roof; field cut 3/16″ fillet weld	0.35	200.0	LF	0.0	LF	0
Casing cover plate-floor; field cut 3/16″ fillet weld	0.35	200.0	LF	0.0	LF	0
SCR/CO						
Seal weld the wall, floor and roof panels	0.35	478.0	LF	0.0	LF	0

2.15.3 Baffles, Modules, Pressure Vessel Sheet 3

Estimate—Baffles	Historical			Estimate		
Description	MH	Quantity	Unit	Quantity	Unit	BM
HRSG—Erect Gas Baffles						**0**
Install side wall gas baffles						
Baffle	0.75	896.0	EA	0.0	EA	0
Fillet weld	0.35	1344.0	LF	0.0	LF	0
Install upper splice gas baffles						
Splice plates	0.75	56.0	EA	0.0	EA	0
U-bolt	0.35	40.0	EA	0.0	EA	0
Install lower gas baffles						
Baffle plates	0.75	90.0	EA	0.0	EA	0
U-bolt connections	0.35	240.0	EA	0.0	EA	0
Header/shipping restraints						
Angle 4×4×3/4×6′-1-5/8″	1.00	80.0	EA	0.0	EA	0
Install the gas baffles to the racks						
Seal plates	0.75	52.0	EA	0.0	EA	0
Fillet weld	0.35	432.0	EA	0.0	EA	0

Estimate—Modules	Historical			Estimate		
HRSG—Modules						**0**
REHTR #2 HP SPHTR	2.20	44.5	ton	0.0	ton	00
#2**HRSG—Modules**						
REHTR #1 HP SPHTR #1	2.20	47.5	ton	0.0	ton	0
HP EVAP IP SPHTR #2 LP	1.30	119.0	ton	0.0	ton	0
SPHTR #2						
HP ECON #2 IP SPHTR #1 LP	1.30	133.5	ton	0.0	ton	0
SPHTR #1 IP EVAP						
HP ECON #1 IP ECON LP EVAP	1.30	118.0	ton	0.0	ton	0
FW HEATER #1	2.50	27.0	ton	0.0	ton	0
FW HEATER #2	2.20	58.5	ton	0.0	ton	0
FW HEATER #2	2.20	58.5	ton	0.0	ton	0

Estimate—Pressure Vessels	Historical			Estimate		
HRSG—Pressure Vessels						
HP steam drum						**0**
Set saddle base plates	24.00	1.0	EA	0.0	EA	0
Set drum	1.40	158.0	ton	0.0	ton	0
IP steam drum						
Set saddle base plates	24.00	1.0	EA	0.0	EA	0
Set drum	2.50	9.0	ton	0.0	ton	0
LP steam drum						
Set saddle base plates	24.00	1.0	EA	0.0	EA	0
Set drum	2.20	30.0	ton	0.0	ton	0
Flash separator						
Set saddle base plates	24.00	1.0	EA	0.0	EA	0
Set drum	2.50	6.0	ton	0.0	ton	0

2.15.4   Platforms, Skids, and SCR/CO Internals Sheet 4

Estimate—Platforms		Historical			Estimate		
Description	MH	Quantity	Unit	Quantity	Unit	BM
HRSG—Steel (Ladders, Platforms, and Grating)						**0**
Install structural steel and platforms						0
Top of HRSG L'×W'	0.40	1890.0	SF	0.0	SF	0
HP, IP, LP drums L'×W'	0.40	216.0	SF	0.0	SF	0
Top of inlet duct L'×W'	0.40	540.0	SF	0.0	SF	0
Right/left side X EA times (L'×W')	0.40	256.0	SF	0.0	SF	0
Stair tower X EA times (L'×W')	0.40	288.0	SF	0.0	SF	0
Misc X EA times(L'×W')	0.40	320.0	SF	0.0	SF	0
Handrail LF	0.32	558.0	LF	0.0	LF	0
Ladders LF	1.00	180.0	LF	0.0	LF	0
Stair tower structural W/silencer	32.00	14.0	ton	0.0	ton	0

Estimate—Skids		Historical			Estimate		
HRSG—Skids						**0**
Level and set the AFCU skid						
Set and level skid	80.00	1.0	EA	0.0	EA	0

Estimate—SCR/CO		Historical			Estimate		
HRSG—SCR and CO$_2$ Internals						**0**
Install Internal Structure						
Remove shipping braces	1.00	50.0	EA	0.0	EA	0
Set support structure and cut/weld, and grind lifting lugs	30.00	1.0	EA	0.0	EA	0
Set left, middle, and right internal structure	30.00	3.0	EA	0.0	EA	0
Weld structure bases to each other	0.35	280.0	LF	0.0	LF	0
Insert and weld top and side pins-4″-1/4″ fillet weld	0.35	9.0	EA	0.0	EA	0
Catalyst Anchor Installation						
Set anchor channel and weld to structure	0.35	108.0	LF	0.0	LF	0
Install T-shaped anchors	0.75	108.0	EA	0.0	EA	0
Make-up T-shaped-washer/nut w/jam nut	0.35	108.0	EA	0.0	EA	0
Seal Plate Installation						
Install side/top/bottom slide plates	2.00	31.0	EA	0.0	EA	0
Weld seal plates	0.35	530.0	LF	0.0	LF	0
Trim seal plates	0.25	132.5	LF	0.0	LF	0

2.15.5   Stack and Breeching Sheet 5

Estimate—Stack and Breeching	Historical			Estimate		
Description	**MH**	**Quantity**	**Unit**	**Quantity**	**Unit**	**BM**
HRSG—Stack and Breeching						**0**
Erect Lower Stack						
Seam Up Shell Cans for Lower Section						
Set shell cans (field fabrication)	4.00	14.0	EA	0.0	EA	0
Weld vertical shell can to shell can	0.75	320.0	LF	0.0	LF	0
Stack Shell Cans and Weld						
Set shell cans	4.00	14.0	EA	0.0	EA	0
Weld horizontal	0.75	350.0	LF	0.0	LF	0
Shim Packs, Erect Lower Stack, and Grout Base Ring						
Shims and bolt to foundation	2.40	28.0	EA	0.0	EA	0
Set lower stack section	60.00	1.0	EA	0.0	EA	0
Erect Damper						
Seam Up (2) Damper Halves						
Set up damper halves (field fabrication)	10.00	2.0	EA	0.0	EA	0
Weld damper halves	0.80	20.0	LF	0.0	LF	0
Bolt up	0.40	100.0	EA	0.0	EA	0
Erect Upper Stack						
Seam Up Shell Cans for Upper Section						
Set shell cans (field fabrication)	4.00	8.0	EA	0.0	EA	0
Weld vertical shell can to shell can	0.75	160.0	LF	0.0	LF	0
Stack Shell Cans and Weld						
Set shell cans	4.00	8.0	EA	0.0	EA	0
Weld horizontal	0.50	200.0	LF	0.0	LF	0
Erect Upper Stack Sections						
Set upper sections	60.00	1.0	EA	0.0	EA	0
Weld horizontal	0.75	50.0	LF	0.0	LF	0
Remove shipping braces	0.25	400.0	EA	0.0	EA	0
Attach Structural Steel and Platforms						
360 degree platform	0.15	360.0	EA	0.0	EA	0
Ladder	0.35	132.0	LF	0.0	LF	0
360 degree platform	0.15	360.0	SF	0.0	SF	0
Bridge	0.15	240.0	SF	0.0	SF	0
Fit and Weld Breech Sections to Stack						
Set panels	60.00	4.0	EA	0.0	EA	0
Weld	0.75	330.0	LF	0.0	LF	0
Install expansion joint						
Expansion joint	60.00	1.0	EA	0.0	EA	0
Backup bars	1.00	24.0	EA	0.0	EA	0
Bolt up	0.25	1120.0	EA	0.0	EA	0
Expanded metal	1.20	200.0	EA	0.0	EA	0
HRSG—Receive, Off Load, and Haul to Hook	978.00	1.0	EA	0.0	EA	0

2.16 HRSG Triple Pressure; Double Wide-Installation Man Hours

Facility-Combined Cycle Power Plant	Actual	Estimated
	MH	Unit
Description	MH	**BM**
HRSG—casing columns	4838	0
HRSG—duct work SCR and inlet	2183	0
Erect—gas baffles	1620	0
HRSG—modules	880	0
HRSG—pressure vessels	420	0
HRSG—steel (ladders, platforms, and grating)	2211	0
HRSG—stack and breeching	2582	0
HRSG—skids	80	0
HRSG—SCR and CO_2 internals	708	0
HRSG—receive, off load and haul to hook	978	0
Equipment installation man hours	**16500**	**0**

2.17   General Scope of Field Work Required for Each Single Wide HRSG

HRSG—Double Pressure—Single Wide
Scope of Work-Field Erection

Base plates/anchor bolts/grout
Set modules/box 1, 2, 3, and 4
Between box 2 and box 3
Between box 3 and Box 4
SCR duct/field joint
Inlet duct/burner/field joint/columns A, B, C, and D
Duct column A-B
Duct column B-C
Duct column C-D
Duct burner assembly
Install main gas and cooling air headers
Flame scanner assembly bolt up to elements
Header assembly
Main gas header:
Cooling air header
Ignition gas header
Instrument air header
Bolt headers to elements
Distribution grid
Assemble grid panels bolt/nut/washer—Tack Weld
Install liner panels
Vessels—drums/deaerator/blow down tank
72″ ID HP steam drum
60″ ID LP steam drum
Deaerator 10′-6″ OD × 27′-7″
Tank 9′-0″ × 5′-0″
Blow down tank 11′-7/8″ × 5′-0″ ID
Skids
Fuel skid
Blower skid
Crossover piping
Pipe rack/DA storage tank silencer
Pipe rack
DA storage tank silencer
HP and LP boiler feed pumps
Platforms
Main deck
Mid level
Lower level
HP and LP drums
Stack
Deaerator and storage tank
Stair tower
Outlet duct/expansion joint/stack
Expansion joint

2.18 HRSG Double Pressure Single Wide Estimate Data

2.18.1 Modules Box 1, 2, 3, 4, Spool Duct and Casing Sheet 1

Description	MH	Unit
Modules-Box 1, 2, 3, 4		
Base Plate	32.00	EA
75,000 lb	2.50	ton
75,001 lb > weight < 125,000 lb	2.20	ton
125,001 lb > weight <= 200,000 lb	1.80	ton
200,001 lb > weight <= 300,000 lb	1.50	ton
300,001 lb > weight <= 400,000 lb	1.40	ton
Box 1, 2, 3, and 4 weld to base plate	1.50	EA
Remove shipping steel box 1, 2, 3, and 4	2.00	EA
Duct transverse top/bottom beam, floor/roof panels; top/bottom cover plate	5.50	EA
Temporary support	16.00	EA
Moment weld	5.50	EA
Bolt bottom and top cover plate to transverse beam	0.15	EA
Weld top and bottom cover plates	0.35	LF
Seal weld panels and bolts	0.35	LF
SCR Duct/Field Joint		
Install floor and roof and side panel	5.50	EA
SCR weld to base plate	10.00	EA
Bolt floor and roof, side panels to columns	0.15	EA
Seal weld panels and bolts	0.35	LF

2.18.2 Inlet Duct, Burner and Distribution Grid Sheet 2

Description	MH	Unit
Inlet Duct/Burner/Field Joint/Columns A, B, C, and D		
Duct column X-X	24.00	EA
Field joint at X bolt and seal weld	0.35	LF
Install insulation and liners	65.00	EA
Inlet duct weld to base plate	5.50	EA
Duct Burner Assembly		
Main gas header	**20.00**	EA
Cooling air header	**18.00**	EA
Ignition gas header	**12.00**	EA
Instrument air header	**20.00**	EA
Bolt headers to elements	**24.00**	EA
Distribution Grid		
Assemble grid panels bolt/nut/washer—tack weld	4.40	EA
Install Access Door		
Splice—grid panel to panel 3/32 fillet weld	0.35	EA
Channel	2.20	EA
Support pipe	2.20	EA
Land grid—strut bolt	2.20	EA
Land grid—weld grid panel end overlap at support pipe	2.20	EA

2.18.3   Liner Panels, Vessels, Skids, Crossover Piping Sheet 3

Description	MH	Unit
Install Liner Panels		
Liner panels/bolt	3.39	EA
Vessels-Drums/Deaerator/Blow Down Tank		
72″ ID HP steam drum	80.00	EA
Drum saddle support	5.50	EA
60″ ID LP steam drum	80.00	EA
Drum saddle support	5.50	EA
Connection to drum		
Deaerator 10′-6″ OD × 27′-7″ 38,903 lb	60.00	EA
Fixed supports	8.80	EA
Tank 9′-0″ × 5′-0″	12.00	EA
Blow down tank 11′-7/8″ × 5′-0″ ID	12.00	EA
Make-up anchor bolts—grout base plates	8.80	EA
Cems	60.00	EA
BMS	80.00	EA
Heat exchanger	24.00	EA
Skids		
Fuel skid	80.00	EA
Blower skid	80.00	EA
Crossover Piping		
Top/bottom		
5″ s120 CS pipe x′-x″	1.54	EA
5″ s120 CS BW	6.25	EA

2.18.4   Pipe Rack, Pumps, Platforms, Duct/Stack Sheet 4

Description	MH	Unit
Pipe Rack/DA Storage Tank Silencer		
Pipe Rack	32.99	ton
DA Storage Tank Silencer		
W4×13 31'	15.95	EA
W6×15 4'	15.95	EA
W4×13 3'	15.95	EA
Bolt up 9	0.35	EA
Base plates 3/16″ fillet weld/grout	2.20	EA
HP and LP Boiler Feed Pumps		
HP feed pump 700 HP set/couple/grout (5″ grout holes)	80.00	EA
HP feed pump 75 hp set/couple/grout (5″ grout holes)	20.00	EA
Platforms		
Main deck	48.00	ton
Mid level	48.00	ton
Lower level	48.00	ton
HP and LP drums	48.00	ton
Stack	48.00	ton
Deaerator and storage tank	48.00	ton
Stair tower	48.00	ton
Outlet Duct/Expansion Joint/Stack		
Stack 10'-1″ diameter×118'-10″		
Section 1 39'-9-7/8″	22.00	EA
Section 2 28'-11-3/8″	22.00	EA
Section 1 48'-5-3/4″	22.00	EA
Field weld horizontal	11.00	EA
Damper		
Inlet flange	11.00	EA
Plate 4'-8-1/2″×52'-9-1/4″	11.00	EA
Plate 4'-8-1/2″×10'-6″		
Bolt up	0.07	EA
Seal weld 3/16″ fillet weld 52'-1″	0.33	LF
Bolt up at foundation AB	0.07	EA
Expansion Joint		
Flex element	22.00	EA
Retaining bars (two sets)		
Fasteners	0.03	EA
Drop wrap gasket	0.02	EA
3/16″ fillet weld	0.33	LF
Hydro	180.00	TEST

2.19 HRSG Double Pressure Single Wide Estimate

2.19.1 Modules Box 1, 2, 3, 4, Spool Duct and Casing Sheet 1

Estimate - Modules Box 1, 2, 3, 4, Spool Duct and Casing	Historical			Estimate		
Description	MH	Quantity	Unit	Quantity	Unit	BM
Modules Box 1, 2, 3, 4, Spool Duct and Casing						**0**
Base plate BP	32.00	5.0	EA	0.0	EA	0
Box #1	1.50	148.0	ton	0.0	ton	0
Box #2	1.50	134.0	ton	0.0	ton	0
Box #3	1.50	124.0	ton	0.0	ton	0
Box #4	1.50	130.0	ton	0.0	ton	0
Box 1, 2, 3, and 4 weld to base plate	1.50	16.0	EA	0.0	EA	0
Remove shipping steel box 1, 2, 3, and 4	2.00	16.0	EA	0.0	EA	0
Spool Duct/Casing/Field Joint/Box 2, 3, and 4						
Between Box 2 and Box 3						
Duct bottom transverse beam	5.50	2.0	EA	0.0	EA	0
Duct top transverse beam	5.50	2.0	EA	0.0	EA	0
Moment weld	5.50	4.0	EA	0.0	EA	0
Install floor and roof panels	5.50	4.0	EA	0.0	EA	0
Temporary support	16.00	1.0	EA	0.0	EA	0
Install bottom and top cover plate	5.50	2.0	EA	0.0	EA	0
Bolt bottom and top cover plate to transverse beam	0.15	76.0	EA	0.0	EA	0
Weld top and bottom cover plates	0.35	88.0	LF	0.0	LF	0
Install side panels						
Bolt side panels to columns and beams	0.15	536.0	EA	0.0	EA	0
Seal weld bolts	0.35	536.0	LF	0.0	LF	0
Seal weld panels	0.35	340.0	LF	0.0	LF	0
Between Box 3 and Box 4						
Duct bottom transverse Beam	5.50	2.0	EA	0.0	EA	0
Duct top transverse beam	5.50	2.0	EA	0.0	EA	0
Moment weld	5.50	4.0	EA	0.0	EA	0
Install floor and roof panels	5.50	4.0	EA	0.0	EA	0
Temporary support	16.00	1.0	EA	0.0	EA	0
Install bottom and top cover plate	5.50	2.0	EA	0.0	EA	0
Bolt bottom and top cover plate to transverse beam	0.15	48.0	EA	0.0	EA	0

Continued

Estimate - Modules Box 1, 2, 3, 4, Spool Duct and Casing		Historical			Estimate		
Description	MH	Quantity	Unit	Quantity	Unit	BM
Seal weld top and bottom cover plates	0.35	88.0	LF	0.0	LF	0
Install side panels						
Bolt side panels to columns and beams	0.15	536.0	EA	0.0	EA	0
Seal weld bolts	0.35	536.0	LF	0.0	LF	0
Seal weld panels	0.35	340.0	LF	0.0	LF	0
SCR Duct/Field Joint						
Install floor and roof	5.50	2.0	EA	0.0	EA	0
Bolt floor and roof to columns	0.15	108.0	EA	0.0	EA	0
Seal weld floor and roof panels	0.35	44.0	LF	0.0	LF	0
Install side panels	5.50	3.0	EA	0.0	EA	0
Bolt side panels to columns	0.15	496.0	EA	0.0	EA	0
Seal weld bolts	0.35	496.0	LF	0.0	LF	0
Seal weld panels	0.35	328.0	LF	0.0	LF	0
SCR weld to base plate	10.00	2.0	EA	0.0	EA	0

2.19.2   Inlet Duct, Burner and Distribution Grid Sheet 2

Estimate - Inlet Duct, Burner and Distribution Grid		Historical		Estimate		
Description	MH	Quantity	Unit	Quantity	Unit	BM
Inlet Duct, Burner and Distribution Grid						**0**
Duct column A-B	24.00	1.0	EA	0.0	EA	0
Field joint at B bolt and seal weld	0.35	55.0	LF	0.0	LF	0
Install insulation and liners	65.00	1.0	EA	0.0	EA	0
Duct column B-C	24.00	1.0	EA	0.0	EA	0
Field joint at C bolt and seal weld	0.35	80.0	LF	0.0	LF	0
Install insulation and liners	65.00	1.0	EA	0.0	EA	0
Duct column C-D	24.00	1.0	EA	0.0	EA	0
Field joint at D bolt and seal weld	0.35	144.0	LF	0.0	LF	0
Install insulation and liners	65.00	1.0	EA	0.0	EA	0
Inlet duct weld to base plate	5.50	6.0	EA	0.0	EA	0
Duct Burner Assembly						
Main Gas, IA, Igniter Gas, and Cooling Air Headers						
Main gas header	**20.00**	1.0	EA	0.0	EA	0
Cooling air header	**18.00**	1.0	EA	0.0	EA	0
Ignition gas header	**12.00**	1.0	EA	0.0	EA	0
Instrument air header	**20.00**	1.0	EA	0.0	EA	0
Bolt headers to elements	**24.00**	1.0	EA	0.0	EA	0
Distribution Grid						
Grid panels bolt/nut/ washer—tack weld	4.40	9.0	EA	0.0	EA	0
Install Access Door						
Splice—grid panel to panel 3/32 fillet weld	0.35	45.0	EA	0.0	EA	0
Channel	2.20	4.0	EA	0.0	EA	0
Support pipe	2.20	4.0	EA	0.0	EA	0
Land grid - strut bolt	2.20	8.0	EA	0.0	EA	0
Weld grid panel end overlap at support pipe	2.20	6.0	EA	0.0	EA	0

2.19.3   Liner Panels, Vessels, Skids, Crossover Piping Sheet 3

Estimate—Liner Panels		Historical		Estimate		
Description	MH	Quantity	Unit	Quantity	Unit	BM
Install Liner Panels						**0**
Liner panels/bolt	3.39	146.0	EA	0.0	EA	0
Liner panels/bolt	3.39	100.0	EA	0.0	EA	0
Liner panels/bolt	3.39	61.0	EA	0.0	EA	0

Estimate—Vessels		Historical		Estimate		
Vessels-Drums/Deaerator/Blow down Tank						**0**
72″ ID HP steam drum	80.00	1.0	EA	0.0	EA	0
Drum saddle support	5.50	1.0	EA	0.0	EA	0
60″ ID LP steam drum	80.00	1.0	EA	0.0	EA	0
Drum saddle support	5.50	1.0	EA	0.0	EA	0
Connection to drum						
Deaerator 10′-6″ OD × 27′-7″ 38,903 lb	60.00	1.0	EA	0.0	EA	0
Fixed supports	8.80	2.0	EA	0.0	EA	0
Tank 9′-0″ × 5′-0″	12.00	1.0	EA	0.0	EA	0
Blow down tank 11′-7/8″ × 5′-0″ ID	12.00	1.0	EA	0.0	EA	0
Make-up anchor bolts—grout base plates	8.80	4.0	EA	0.0	EA	0
Cems	60.00	1.0	EA	0.0	EA	0
BMS	80.00	1.0	EA	0.0	EA	0
Heat exchanger	24.00	1.0	EA	0.0	EA	0

Estimate—Skids		Historical		Estimate		
Skids						**0**
Fuel skid	80.00	1.0	EA	0.0	EA	0
Blower skid	80.00	1.0	EA	0.0	EA	0

Estimate—Crossover Piping		Historical		Estimate		
Crossover Piping						**0**
Top						
5″ s120 CS Pipe x′-x″	1.54	7.0	EA	0.0	EA	0
5″ s120 CS BW	6.25	4.0	EA	0.0	EA	0
Bottom						
5″ s120 CS Pipe 3′-3″	1.54	4.0	EA	0.0	EA	0
5″ s120 CS BW	6.25	2.0	EA	0.0	EA	0

2.19.4 Pipe Rack, Pumps, Platforms, Duct/Stack Sheet 4

Estimate—Pipe Rack, DA Storage Tank Silencer		Historical		Estimate		
Description	MH	Quantity	Unit	Quantity	Unit	BM
Pipe Rack/DA Storage Tank Silencer						**0**
Pipe rack	41.30	**7.8**	ton	0.0	ton	0
DA Storage Tank Silencer						
W4×13 31′	24.00	0.8	ton	0.0	ton	0
W6×15 4′	24.00	0.2	ton	0.0	ton	0
W4×13 3′	24.00	0.2	ton	0.0	ton	0
Bolt up 9	0.35	180.0	EA	0.0	EA	0
Base plates 3/16″ Fillet weld/grout	4.00	4.0	EA	0.0	EA	0

Estimate—Pumps		Historical		Estimate		
HP and LP Boiler Feed Pumps						**0**
HP feed pump 700 hp set/couple/grout	80.00	3.0	EA	0.0	EA	0

Estimate—Platforms		Historical		Estimate		
Platforms						**0**
Main deck	48.00	**6.4**	ton	0.0	ton	0
Mid level	48.00	**1.9**	ton	0.0	ton	0
Lower level	48.00	**2.9**	ton	0.0	ton	0
HP and LP drums	48.00	**1.0**	ton	0.0	ton	0
Stack	48.00	**1.1**	ton	0.0	ton	0
Deaerator and storage tank	48.00	**3.2**	ton	0.0	ton	0
Stair tower	48.00	**13.2**	ton	0.0	ton	0

Estimate—Outlet Duct, Stack		Historical		Estimate		
Outlet Duct/Expansion Joint/Stack						**0**
Stack 10′-1′′ diameter × 118′-10″						
Section 1 39′- 9-7/8″	22.00	1.0	EA	0.0	EA	0
Section 2 28′- 11-3/8″	22.00	1.0	EA	0.0	EA	0
Section 1 48′- 5-3/4″	22.00	3.0	EA	0.0	EA	0
Field weld horizontal	12.00	1.0	EA	0.0	EA	0
Damper						
Inlet flange	11.00	2.0	EA	0.0	EA	0
Plate 4′-8-1/2′′×52′-9-1/4″	11.00	2.0	EA	0.0	EA	0
Plate 4′-8-1/2′′×10′-6″						
Bolt up	0.15	224.0	EA	0.0	EA	0
Seal weld 3/16″ fillet weld 52′-1″	0.35	125.0	LF	0.0	LF	0
Bolt up at foundation AB	0.15	28.0	EA	0.0	EA	0
Expansion Joint						
Flex element	22.00	1.0	EA	0.0	EA	0
Retaining bars (two sets)						
Fasteners	0.03	1004.0	EA	0.0	EA	0
Drop wrap gasket	0.02	460.0	EA	0.0	EA	0
3/16″ fillet weld	0.35	125.0	LF	0.0	LF	0
Hydro	180.00	1.0	TEST	0.0	TEST	0

2.20   HRSG Double Pressure; Single Wide-Installation Man Hours

Facility-Combined Cycle Power Plant	Actual	Estimated
Description	MH	BM
Modules box 1, 2, 3, 4, spool duct and casing	2506	0
Inlet duct, burner, and distribution grid	595	0
Install liner panels	1040	0
Vessels—drums/deaerator/blowdown tank	472	0
Skids	160	0
Crossover piping	54	0
Pipe rack/DA storage tank silencer	428	0
HP and LP boiler feed pumps	240	0
Platforms	1423	0
Outlet duct/expansion joint/stack	537	0
Equipment installation man hours	**7456**	**0**

2.21 General Scope of Field Work Required for Each Single Pressure MAWP HRSG

Boiler Typical:

Single pressure MAWP = 126.4 bsrg/1834 psig
Double wide gas path
Bundle design (4 modules)
CO and SCR
Drum plus design (eight separator bottles)
Heavy earthquake design
Stack height 44.5–67 m/146–220 ft
Stack diameter 6.5 m (inside)
Silencer in stack above stack damper

Field Scope of Work

Main Steel—Preassemble and set goal post
Boiler casing
Place casing and bolt case panels to main steel (goal post)
Weld case panels (outside seam)
Weld case panels (inside stitch)
Install inlet casing
Install insulation and liners
Install separator bottles
Install all bracing
Modules offload, set and align
Install drum
Install SCR support structure
Load SCR
Install platforms and stairway
Install internal AIG sections
Install gas baffles; boiler/inlet duct
Install stack, damper, silencer, breeching, and outlet duct
Install stack dampers and silencer
Install stack platforms and caged ladders
Personnel protection
Hydro test

2.22 HRSG Single Pressure Double Wide Estimate Data

2.22.1 Main Steel, Boiler and Inlet Casing and Liners Sheet 1

Description	MH	Unit
Main Steel—Preassemble and Set Goal Post		
Erect goal post and place bracing	15.60	ton
Erect structural steel frame at column	14.31	ton
Boiler Casing		
Place casing and bolt case panels to main steel (goal post)		
Place and bolt side panels to goal post	10.00	EA
Place outlet flange OF1, OF3, OF4, OF6	6.00	EA
Place outlet flange OF2, OF5, T6, B4	8.00	EA
Bolt case panels to goal post	0.10	EA
Weld case panels	0.20	LF
Install Inlet Casing		
Place side panels	20.00	EA
Bolt case panels to goal post	0.11	EA
Weld inlet duct panels	0.20	LF
Install Insulation and Liners		
Install liners	1.50	EA
Insulation	220.50	EA

2.22.2 Separator Bottles, Bracing, Modules, SCR and Platforms and Stairway Sheet 2

Description	MH	Unit
Install Separator Bottles	40.00	EA
Install All Bracing		
Boiler casing:		
Shoulder support SH1–SH15	4.00	EA
Bolt connection	0.27	EA
Loose plate P1-P14	2.00	EA
Inlet duct; loose plate and assemblies	4.00	EA
Modules Offload, Set and Align		
Rig and set boxes	3.28	ton
Shipping frame	40.00	EA
Install Drum		
HP drum	4.10	ton
Install SCR Support Structure		
Column, beams, and cross bracing	20.00	EA
Load SCR	640	Unit
Install Platforms and Stairway		
Install platforms	0.45	SF
Stair Tower:		
Column	20.00	EA
Beam	6.00	EA
Stairs	12.25	EA
Handrail	0.25	LF

2.22.3 AIG, Baffles, Stack and Platforms, Personnel Protection, Hydro Test Sheet 3

Description	MH	Unit
Install Internal AIG Sections		
Assemble sections	80	EA
Hoist sections from top HRSG casing	60	EA
Fix AIG grid in casing wall	140	EA
Field weld joints	0.582	LF
Close roof all field joints	460	JOB
Install Gas Baffles; Boiler/Inlet Duct		
Pipe and transverse bumpers	20.00	EA
Bumper support guide	10.00	EA
Baffles:		
Sealing plates CO/SCR	8.01	EA
Harp side wall gas baffles boxes	8.00	EA
Install Stack, Damper, Silencer, Breeching, and Outlet Duct		
Stack sections (half sections)	60.00	ton
Weld vertical seam	1.29	LF
Field splice-weld horizontal seam	1.15	LF
Bolt lower section to foundation	60.00	JOB
Set stack sections	40.00	ton
Outlet duct:		
W14×120	12.82	ton
C10×15.3	70.66	ton
L8×8×3/4	10.00	EA
Angle braces	2.00	EA
Boiler and duct outlet flange	80.00	EA
Bolted connection	1.18	EA
Install Stack Dampers and Silencer		
Stack damper; set and weld	60	JOB
Silencer	60	EA
Install Stack Platforms and Caged Ladders		
Stack platforms	0.11	SF
Handrail	0.30	LF
Caged ladders	0.5	LF
Personnel Protection	480	JOB
Hydro Test	960	TEST

2.23   HRSG-Single Pressure, Double Wide Estimate

2.23.1   Main Steel, Boiler and Inlet Casing and Liners Sheet 1

Estimate - Main Steel	Historical			Estimate		
Description	MH	Quantity	Unit	Quantity	Unit	BM
Main Steel—						**0**
Preassemble and						
Set Goal Post						
Goal post assembly on ground						
Erect goal post and place bracing	15.6	**14.4**	ton	0.0	ton	0
Structural steel at El. 95'-0"						
Column line A and L; frame at column 10	14.3	**15.7**	ton	0.0	ton	0
Column line A and L; frame at column 9	14.3	**31.0**	ton	0.0	ton	0
Column line A and L; frame at column 6	14.3	**15.7**	ton	0.0	ton	0
Column line A and L; frame at column 5	14.3	**27.7**	ton	0.0	ton	0
Column line A and L; frame at column 7	14.3	**8.6**	ton	0.0	ton	0
Column line A and L; frame at column 4	14.3	**3.8**	ton	0.0	ton	0
Column line A and L; frame at column 3	14.3	**3.4**	ton	0.0	ton	0

Estimate—Boiler Casing		Historical		Estimate		
Boiler Casing						**0**
Place Casing and Bolt Case Panels to Main Steel						
Place right side panel R1-R23	10.0	46.0	EA	0.0	EA	0
Place left side panel L1-L23	10.0	46.0	EA	0.0	EA	0
Place outlet flange OF1, OF3, OF4, OF6	6.0	32.0	EA	0.0	EA	0
Place outlet flange OF2, OF5, T6, B4	8.0	8.0	EA	0.0	EA	0
Bolt case panels to goal post	0.10	1008.0	EA	0.0	EA	0
Weld Case Panels (Outside Seam)	0.20	16750.5	LF	0.0	LF	0
Weld Case Panels (Inside Stitch)	0.20	8498.0	LF	0.0	LF	0

Estimate—Inlet Casing	Historical			Estimate		
Install Inlet Casing						**0**
Place top side panel TP01-TP15	20.0	30.0	EA	0.0	EA	0
Place bottom side panel BP01-BP09	20.0	30.0	EA	0.0	EA	0
Place left side panel LP01-LP13	20.0	26.0	EA	0.0	EA	0
Place left side panel RP01-RP13	20.0	26.0	EA	0.0	EA	0
Bolt case panels to goal post	0.11	**2240.0**	EA	0.0	EA	0
Weld inlet duct panels	0.20	4041.0	LF	0.0	LF	0

Estimate - Liners	Historical			Estimate		
Install Insulation and Liners						**0**
Part 1 inlet duct top liner	1.5	64.0	EA	0.0	EA	0
Part 2 inlet duct top liner	1.5	111.0	EA	0.0	EA	0
Inlet duct front liner	1.5	30.0	EA	0.0	EA	0
Inlet duct right side wall liner	1.5	79.0	EA	0.0	EA	0
Inlet duct left side wall liner	1.5	79.0	EA	0.0	EA	0
Inlet duct bottom liner	1.5	92.0	EA	0.0	EA	0
Boiler top box 1 liner	1.5	161.0	EA	0.0	EA	0
Top SCR box and box 2 liner	1.5	140.0	EA	0.0	EA	0
Boiler right side liner	1.5	195.0	EA	0.0	EA	0
Boiler left side liner	1.5	193.0	EA	0.0	EA	0
Boiler bottom box liner	1.5	58.0	EA	0.0	EA	0
Boiler bottom SCR box and box 2	1.5	171.0	EA	0.0	EA	0
Insulation	220.5	1.0	EA	0.0	EA	0

2.23.2 Separator Bottles, Bracing, Modules, SCR, and Platforms and Stairway Sheet 2

Estimate—Separator Bottles		Historical		Estimate		
Description	**MH**	**Quantity**	**Unit**	**Quantity**	**Unit**	**BM**
Install Separator Bottles	40.00	8.0	EA	0.0	EA	**0**

Estimate—Bracing		Historical		Estimate		
Install Bracing						**0**
Boiler casing:						
Shoulder support SH1-SH15	4.00	100.0	EA	0.0	EA	0
Bolt connection	0.27	90.0	EA	0.0	EA	0
Loose plate P1-P14	2.00	402.0	EA	0.0	EA	0
Inlet duct; loose plate and assemblies	4.00	208.0	EA	0.0	EA	0

Estimate—Modules		Historical		Estimate		
Modules Offload, Set and Align						**0**
Right/left box 1	3.28	280.4	ton	0.0	ton	0
Right/left box 2	3.28	280.4	ton	0.0	ton	0
Shipping frame	40.00	4.0	EA	0.0	EA	0
Install Drum						
HP drum	4.10	34.2	ton	0.0	ton	0

Estimate—SCR		Historical		Estimate		
Install SCR Support Structure						**0**
Column, beams and cross bracing	20.00	38.0	EA	0.0	EA	0
Load SCR	640.00	1.0	Unit	0.0	Unit	0

Estimate—Platform and Stairway		Historical		Estimate		
Install Platforms and Stairway						**0**
Platform at El 95'-8-1/4"	0.45	699.5	SF	0.0	SF	0
Steam drum platform at El 102'-6-15/16"	0.45	90.0	SF	0.0	SF	0
Silencer support structure	0.45	28.0	SF	0.0	SF	0
Top casing platform	0.45	558.8	SF	0.0	SF	0
Platform at El 5'-2-15/16"	0.45	115.0	SF	0.0	SF	0
Platform at El 8'-5-1/8"	0.45	42.0	SF	0.0	SF	0
Platform at El 24'-5-3/16"	0.45	24.0	SF	0.0	SF	0
Platform at El 71'-2-1/4"	0.45	24.0	SF	0.0	SF	0
Platform at El 78'-0"	0.45	138.0	SF	0.0	SF	0
Platform at El 92'-6"	0.45	84.0	SF	0.0	SF	0
Stair tower:						
Column	20.00	6.0	EA	0.0	EA	0
Beam	6.00	36.0	EA	0.0	EA	0
Stairs	12.25	8.0	EA	0.0	EA	0
Handrail	0.25	200.0	LF	0.0	LF	0
Platform	0.45	144.0	SF	0.0	SF	0

2.23.3 AIG, Baffles, Stack and Platforms, Personnel Protection, Hydro Test Sheet 3

Estimate—AIG		Historical		Estimate		
Description	MH	Quantity	Unit	Quantity	Unit	BM
Install Internal AIG Sections						**0**
Assemble sections	80	4.0	EA	0.0	EA	0
Hoist sections from top HRSG casing	60	4.0	EA	0.0	EA	0
Fix AIG grid in casing wall	140	4.0	EA	0.0	EA	0
Field weld joints	0.58	378.0	LF	0.0	LF	0
Close roof all field joints	460	1.0	JOB	0.0	JOB	0

Estimate—Baffles		Historical		Estimate		
Install Gas Baffles; Boiler/ Inlet Duct						**0**
Pipe bumpers	20.00	7.0	EA	0.0	EA	0
Transverse bumpers IA1, IB1, IB2, IC1	20.00	4.0	EA	0.0	EA	0
Bumper support guide BSA1, BSA2	10.00	2.0	EA	0.0	EA	0
Baffles:						
Sealing plates CO	8.01	85.0	EA	0.0	EA	0
Sealing plates SCR	8.01	72.0	EA	0.0	EA	0
Harp side wall gas baffles box 1	8.00	102.0	EA	0.0	EA	0
Harp side wall gas baffles box 2	8.00	148.0	EA	0.0	EA	0

Estimate—Stack		Historical		Estimate		
Install Stack, Damper, Silencer, Breeching, Outlet Duct						**0**
Stack sections (half sections)	60.00	12.0	ton	0.0	ton	0
Weld vertical seam	1.29	420.0	LF	0.0	LF	0
Field splice-weld horizontal seam	1.15	540.1	LF	0.0	LF	0
Bolt lower section to foundation	60.00	1.0	JOB	0.0	JOB	0
Set stack sections	80.00	6.0	ton	0.0	ton	0
Outlet duct:						
W14×120	12.8	9.4	ton	0.0	ton	0
C10×15.3	70.7	0.3	ton	0.0	ton	0
L8×8×3/4	10.00	2.0	EA	0.0	EA	0
Angle braces	2.00	20.0	EA	0.0	EA	0

Continued

Estimate—Stack	Historical			Estimate		
Expansion joint:						
Boiler outlet flange	80.00	1.0	EA	0.0	EA	0
Outlet duct flange	80.00	1.0	EA	0.0	EA	0
Bolted connection	1.18	68.0	EA	0.0	EA	0
Install Stack Dampers and Silencer						
Stack damper; set and weld	60	1.0	JOB	0.0	JOB	0
Silencer	60	1.0	EA	0.0	EA	0

Estimate—Stack Platforms	Historical			Estimate		
Install Stack Platforms and Caged Ladders						**0**
Stack platforms	0.11	1080.2	SF	0.0	SF	0
Handrail	0.30	270.0	LF	0.0	LF	0
Caged ladders	0.5	120.0	LF	0.0	LF	0

Estimate—Personnel Protection	Historical			Estimate		
Personnel protection	480	1.0	JOB	0.0	JOB	0

Estimate—Hydro Test	Historical			Estimate		
Hydro test	960	1.0	TEST	0.0	TEST	0

2.24 HRSG Single Pressure; Double Wide-Installation Man Hours

	Actual	Estimated
Facility-Combined Cycle Power Plant	**MH**	**BM**
Main steel—preassemble and set goal post	1740	0
Boiler casing	6379	0
Install inlet casing	3300	0
Install insulation and liners	2280	0
Install separator bottles	320	0
Install bracing	2060	0
Modules offload, set and align	2140	0
Install SCR support structure	1400	0
Install platforms and stairway	1360	0
Install internal AIG sections	1800	0
Install gas baffles; boiler/inlet duct	3498	0
Install stack, damper, silencer, breeching, outlet duct	2980	0
Install stack platforms and caged ladders	260	0
Personnel protection	480	0
Hydro test	960	0
Equipment installation man hours	**30958**	**0**

2.25   General Scope of Field Work Required for Each Air Cooled Condenser

Scope of Work-Field Erection

Off load and store ACC equipment (transport material to work area)
Erect structural steel including stairs and access Platforms
Erect structural steel modules 1–6
Fan duct support steel modules 1–6
Preassemble stairways at grade and set and secure
Preassemble external walkways at grade and set and bolt up
Upper structure A-frame steelwork
Field assemble fan deck including fans, gearbox's and fan cowls
Fan inlet cowl and fan screens
Motor support bridge
Erect and install wind wall
Erect and weld tube bundles
Erect and weld condensate headers
Field assemble street steam duct
Erect and weld interconnecting street pipe
Erect weld interconnecting equipment pipe
Field assemble transition duct to steam turbine
Field assemble street steam duct
Field assemble risers
Erect and weld out risers
Set and assemble shop tanks
Set and assemble pumps
Erect and weld out transition to steam turbine
Set and connect steam jet air ejectors
Pneumatic testing

2.26   Air Cooled Condenser Estimate Data

2.26.1   Structural Steel Module/Fan Duct, Stairways and Walkways, A-Frame and Fan Cowl/Screens Sheet 1

Description	MH	Unit
Erect and temporarily bolt structural steel	2.50	ton
Bolt connection	0.15	EA
Moment weld	4.00	EA
Handle and place transverse	5.00	EA
Base plate	2.00	EA
Grout base plate	1.40	SF
Preassemble Stairways at Grade and Set and Secure		
Unload, assemble, and install frame	0.10	SF
Install handrail	0.20	LF
Install stairs	0.85	LF
Install grading	0.17	SF
Preassemble External Walkways at Grade and Set and Bolt Up		
Preassemble walkways, set and bolt	0.20	SF
Upper Structure A-Frame Steelwork		
Seal plate	2.00	EA
Access door frame	10.00	EA
Cut sheeting to size	0.002	SF
Place sheeting	0.23	SF
Fix sheeting w/flashing and caulk	0.05	SF
Tack or screw sheeting	1.00	EA
Install access door	10.00	EA
Pipe hanger w/turnbuckle and clevis	10.00	EA
Install steam saddles	10.00	EA
Bolt connection	0.15	EA
Fan Inlet Cowl and Fan Screens		
Fan screen supports	2.0	EA
Fan screen modules	3.0	EA
Mount fan cowl segments, stiffening plates, and top angle ring	2.0	EA
Install fan deck plates	2.0	EA

2.26.2 Motor Bridge, Tube Bundles, Condensate Headers, and Steam Distribution Manifolds Sheet 2

Description	MH	Unit
Motor Support Bridge		
20″ plate	2.2	EA
Place fan motor and gearbox	4.0	EA
Bolt fan hub	4.0	EA
Install fan blades	3.0	EA
Lift assembly onto fan deck	4.0	EA
Bolt bridge to structure	4.0	EA
Erect Tube Bundles		
Remove outlet/inlet tube sheet plate	4.0	EA
Unbolt adjusting plates and install and air seals	4.0	EA
Fix bundle keeper and secondary bundle air seals	4.0	EA
Lift and set tube bundles and tack to CCM bundles	12.0	EA
Weld Tube Bundles		
Weld inlet tube sheets together (0.04″ BW)	0.60	LF
Weld condensate manifold to outlet tube plates	0.75	LF
Weld CCM to lower tube sheets (0.79″ BW)	1.20	LF
Erect Condensate Headers		
Install condensate headers	10.0	EA
Install saddles	2.0	EA
Bolt connection to A frame	2.3	EA
Weld Condensate Headers		
16″ schedule 10 BW CS	12.0	JT
18″ schedule 10 BW CS	13.5	JT
Erection of Steam Distribution Manifolds		
Handle/set duct	5.6	EA
Install platform	6.0	EA
Install fixed and sliding saddles	1.0	EA
Make-up stiffener rings	2.0	EA
Movable ladder	2.0	EA
Install end plate	4.0	EA
Weld Steam Distribution Manifolds		
6.5′–8.5′ diameter × 5/16″ FW	0.30	LF

2.26.3 IC Equipment and Street Pipe, Wind Walls, Turbine Exhaust Duct Sheet 3

Description	MH	Unit
Erect and Weld IC Equipment and Street Pipe		
Handle pipe	0.050	DI
STD BW	0.50	DI
Instrument	2.00	EA
Valves	0.65	DI
Pipe support	1.0	DI
Hydro	0.11	LF
Erection and Sheeting of Wind Walls		
Place sheeting	0.23	SF
Fix sheeting w/flashing and caulk	0.05	SF
Tack or screw sheeting	1.00	EA
Erect and Weld Turbine Exhaust Duct		
Handle 48″ diameter duct	10.00	EA
Weld duct; 48″ diameter × 0.75″ WT BW	1.15	LF
48″ diameter manhole	1.7	DI
Weld in nozzle w/reinforcement pad	2.25	LF
Install and align embedded plates	8.0	EA
Weld together saddles and weld stiffeners	4.0	EA
Install saddles	4.0	EA
Bolt saddles to foundation	2.0	EA
Place duct sections	10.0	EA
Grout support saddles	8.0	EA
Weld vanes into elbow		
Install partition plate	80.0	EA
Weld partition plate 0.86″ BW	1.15	LF
Install vanes	4	EA
Weld vanes fillet weld	2	EA
Cut lifting lug	2	EA
Instrumentation	4	EA
Remove blank plate	0.15	EA
Install stiffener plate	0.6	EA
10″ s80 BW	7.5	EA
10″ s 80 fillet weld	7.5	EA

2.26.4   Street Steam Duct Sheet 4

Description	MH	Unit
Field Assemble Street Steam Duct		
Handle 20'-6" diameter duct	10.00	EA
Handle drain pot	10.00	EA
Erect platform	10.00	EA
Field weld w/slip ring 0.75" BW	1.15	LF
Weld duct; 20'-6" diameter × 0.75" WT BW	1.15	LF
Fit up and weld 0.75" BW	1.15	LF
Install and align embedded plates	8.00	EA
Weld together saddles and weld stiffeners	4.00	EA
Install saddles	4.00	EA
Bolt saddles to foundation	2.00	EA
Place duct sections	10.00	EA
TED-XX handle duct	10.00	EA
Grout support saddles	8.00	EA
Reinforcement Pad	1.15	LF

2.26.5   Erect and Weld Turbine Exhaust Duct Risers Sheet 5

Description	MH	Unit
Erection and Welding of Turbine Exhaust Duct Risers		
Erect duct section, duct w/elbow	10.00	EA
Handle duct section, duct section w/elbow	10.00	EA
Fit up and weld 0.75" BW		
FFW	1.15	LF
FW	1.15	LF
Install and align embedded plates	8.00	EA
Weld together saddles and weld stiffeners	4.00	EA
Install saddles	4.00	EA
Bolt saddles to foundation	2.00	EA

2.26.6   Risers, Shop Tanks Sheet 6

Description	MH	Unit
Field Assemble Risers		
Handle duct	10.0	EA
Expansion joint 8'-6"	10.0	EA
Set and Assemble Shop Tanks		
Set condensate tank	40.0	EA
Set deaerator tank		
3'-11" diameter	20.0	EA
8'-6" diameter	20.0	EA
Fit up and weld		
0.5"×25	0.75	LF
0.5"×54	0.75	LF
Perforated plate	20.0	EA
Wire mesh	20.0	EA

2.26.7   Pumps, Expansion Joint, Test Sheet

Description	MH	Unit
Pumps		
Set pump skid and pumps	120.0	EA
Erection and Welding of Expansion Joint (Dog Bone) at Turbine Connection		
Weld together upper dog bone holder (including internal braces)		
Handle holder plates	4.0	EA
Handle corner section	4.0	EA
Fit up and weld holder plates 1.8" thick	4.0	EA
Fit up and weld corner plates 1.1/16" thick and 7/8" thick	4.2	LF
Internal braces 5" diameter std pipe	4.0	EA
Fillet weld	2.0	EA
Set upper dog bone to steam turbine duct	40.0	EA
Fit up and weld to SPX connection 3/4" thick	1.1	LF
Fit up and weld SPX to turbine exhaust duct 2 1/4" thick	2.5	LF
Install dog bone templates	2.0	EA
Rod bar	2.0	EA
Remove steel dog bone templates	2.0	EA
Install rubber at dog bone	1.0	EA
Clamping piece splice	1.0	EA
Field erection bolts	0.2	EA
Pneumatic Testing		
Pneumatic testing	1200	TEST

2.27 Air Cooled Condenser Installation Estimate

2.27.1 Structural Steel Module/Fan Duct, Stairways and Walkways, A-Frame, and Fan Cowl/Screens Sheet 1

Description	MH	Historical Quantity	Unit	Estimate Quantity	Unit	BM	IW	MW
Erect Module Structural Steel						0	0	0
Erect and temporarily bolt structural steel	2.50	11.26	ton	0	ton		0	
Bolt connection	0.15	84.00	EA	0	EA		0	
Moment weld	4.00	8.00	EA	0	EA		0	
Handle and place transverse	5.00	24.00	EA	0	EA		0	
Base plate	2.00	10.00	EA	0	EA		0	
Grout base plate	1.40	36.00	SF	0	SF		0	
Fan Duct Support Steel Module						0	0	0
Erect and temporarily bolt structural steel	2.50	1.50	ton	0	ton		0	
Bolt connection	0.15	1.33	EA	0	EA		0	
Preassemble Stairways at Grade and Set						0	0	0
Unload, assemble and install frame	0.10	19.94	SF	0	SF		0	
Install handrail	0.20	8.39	LF	0	LF		0	
Install stairs	0.85	3.78	LF	0	LF		0	
Install grading	0.17	15.67	SF	0	SF		0	
Erect and temporarily bolt structural steel	2.50	0.42	ton	0	ton		0	
Grout base plate	1.40	0.22	SF	0	SF		0	
Bolt connection	0.15	2.53	EA	0	EA		0	
Base plate	2.00	0.06	EA	0	EA		0	
Preassemble Walkways at Grade and Set						0	0	0
Preassemble walkways, set and bolt	0.20	94.37	SF	0	SF		0	
Upper Structure A-Frame Steelwork						0	0	0
Erect and	2.50	8.52	ton	0	ton		0	
temporarily bolt structural steel								
Seal plate	2.00	20.00	EA	0	EA		0	
Access door frame	10.00	1.00	EA	0	EA		0	
Cut sheeting to size	0.002	626.00	SF	0	SF		0	
Place sheeting	0.23	420.00	SF	0	SF		0	
Fix sheeting	0.05	626.00	SF	0	SF		0	
w/flashing and caulk								
Tack or screw sheeting	1.00	25.00	EA	0	EA		0	
Install access door	10.00	1.00	EA	0	EA		0	
Pipe hanger	10.00	2.00	EA	0	EA		0	
w/turnbuckle and clevis								
Install steam saddles	10.00	3.00	EA	0	EA		0	
Bolt connection	0.15	74.00	EA	0	EA		0	
Fan Inlet Cowl and Fan Screens						0	0	0
Fan screen supports	2.00	12.00	EA	0	EA		0	
Fan screen modules	3.00	5.00	EA	0	EA		0	
Mount fan cowl segments and top ring	2.00	12.00	EA	0	EA		0	
Bolt connection	0.15	36.00	EA	0	EA		0	
Install fan deck plates	2.00	12.00	EA	0	EA		0	

2.27.2 Motor Bridge, Tube Bundles, Condensate Headers, and Steam Distribution Manifolds Sheet 2

Description	MH	Historical Quantity	Unit	Estimate Quantity	Unit	BM	IW	MW
Motor Support Bridge						**0**	**0**	**0**
20″ plate	2.21	6.00	EA	0	EA		0	
Place fan motor and gearbox	4.00	1.00	EA	0	EA		0	
Bolt fan hub	4.00	1.00	EA	0	EA		0	
Install fan blades	3.00	5.00	EA	0	EA		0	
Lift assembly onto fan deck	4.00	1.00	EA	0	EA		0	
Bolt bridge to structure	4.00	1.00	EA	0	EA		0	
Bolt connection	0.15	31.00	EA	0	EA		0	
Handrail	0.20	8.00	LF	0	LF		0	
Stairs	0.85	8.00	EA	0	EA		0	
Erect and	2.50	4.53	ton	0	ton		0	
temporarily bolt structural steel								
Grating	0.17	100.00	SF	0	SF		0	
Erect and Weld Tube Bundles						**0**	**0**	**0**
Erect Tube Bundles						**0**	**0**	**0**
Remove outlet/inlet tube sheet plate	4.00	2.00	EA	0	EA	0		
Unbolt plates and install and air seals	4.00	2.00	EA	0	EA	0		
Bundle keeper and secondary bundle	4.00	2.00	EA	0	EA	0		
air seals								
Lift and set tube bundles	12.00	2.00	EA	0	EA	0		
Weld Tube Bundles						**0**	**0**	**0**
Weld inlet tube sheets together (0.04″	0.60	42.00	LF	0	LF	0		
BW)								
Remove inlet tube plate protection	4.00	2.00	EA	0	EA	0		
sheets								
Weld condensate manifold	0.75	42.00	LF	0	LF	0		
Weld CCM to lower tube sheets (0.79″	1.20	42.00	LF	0	LF	0		
BW)								
Erect and Weld Condensate Headers						**0**	**0**	**0**
Erect Condensate Headers						**0**	**0**	**0**
Install condensate headers	10.00	7.00	EA	0	EA	0		
Install saddles	2.00	13.00	EA	0	EA	0		
Bolt connection to A frame	2.30	13.00	EA	0	EA	0		
Weld Condensate Headers						**0**	**0**	**0**
16″ schedule 10 BW CS	12.00	2.00	JT	0	JT	0		
18″ schedule 10 BW CS	13.50	4.00	JT	0	JT	0		
Erect and Weld Steam Distribution						**0**	**0**	**0**
Manifolds								
Erection of Steam Distribution						**0**	**0**	**0**
Manifolds								
Handle/set duct	5.60	7.00	EA	0	EA			0
Install platform	6.00	1.00	EA	0	EA			0
Install fixed and sliding saddles	1.00	7.00	EA	0	EA			0
Make-up stiffener rings	2.00	7.00	EA	0	EA			0
Movable ladder	2.00	6.00	EA	0	EA			0
Install end plate	4.00	1.00	EA	0	EA			0
Weld Steam Distribution Manifolds						**0**	**0**	**0**
6.5′–8.5′ diameter×5/16″ FW	368.82	0.30	LF	0	LF			0

2.27.3 IC Equipment and Street Pipe, Wind Walls, Turbine Exhaust Duct Sheet 3

Description	Historical			Estimate				
	MH	Quantity	Unit	Quantity	Unit	BM	IW	MW
Erect and Weld IC Equipment Pipe						0	0	0
Handle pipe	0.05	456.7	DI	0	DI			0
Standard BW	0.50	49.3	DI	0	DI			0
Instrument	2.00	0.4	EA	0	EA			0
Valves	0.65	0.8	DI	0	DI			0
Pipe support	1.00	21.8	DI	0	DI			0
Hydro	0.11	99.1	LF	0	LF			0
Erect and Weld IC Street Pipe						0	0	0
Handle pipe	0.05	1288.0	DI	0	DI			0
Standard BW	0.50	76.0	DI	0	DI			0
Pipe support	1.00	43.3	DI	0	DI			0
Hydro	0.11	90.0	LF	0	LF			0
Erection and Sheeting of Wind Walls						0	0	0
Erect and temporarily bolt structural steel	2.50	7.9	ton	0	ton		0	
Door frame	10.00	3.0	EA	0	EA		0	
Bolt connection A-frame	0.15	10.0	EA	0	EA		0	
Bolt connection	0.15	154.0	EA	0	EA		0	
Place sheeting	0.23	260.0	SF	0	SF		0	
Fix sheeting w/flashing and caulk	0.05	480.0	SF	0	SF		0	
Tack or screw sheeting	1.00	16.0	EA	0	EA		0	
Erect and Weld Turbine Exhaust Duct						0	0	0
Handle 48″ diameter duct	10.00	0.9	EA	0	EA	0		
Weld duct; 48′′ diameter × 0.75″ WT BW	1.15	5.9	LF	0	LF	0		
48″ diameter manhole	1.70	1.4	DI	0	DI	0		
Weld in nozzle w/reinforcement pad	2.25	4.9	LF	0	LF	0		
Install and align embedded plates	8.00	0.1	EA	0	EA	0		
Weld together saddles and weld stiffeners	4.00	1.3	EA	0	EA	0		
Install saddles	4.00	0.2	EA	0	EA	0		
Bolt saddles to foundation	2.00	0.2	EA	0	EA	0		
Place duct sections	10.00	0.2	EA	0	EA	0		
Fit up and weld duct sections 0.75″ BW	1.15	32.0	LF	0	LF	0		
Grout support saddles	8.00	0.2	EA	0	EA	0		
Handle 48″ diameter duct	10.00	1.4	EA	0	EA	0		
Fit up and weld sections together								
TED 16, 17 0.75″ BW	1.15	3.0	LF	0	LF	0		
Place duct sections	10.00	0.3	EA	0	EA	0		
Fit up and weld								
TED 16, 17 0.75″ BW	1.15	28.5	LF	0	LF	0		
TED 17, 15, 7A, 7B 0.75″ BW	1.15	11.3	LF	0	LF	0		
Install partition plate	4.00	2.2	EA	0	EA	0		
Weld partition plate 0.86″ BW	44.88	0.0	LF	0	LF	0		
Install vanes	18.00	0.1	EA	0	EA	0		
Weld vanes fillet weld	54.00	0.1	EA	0	EA	0		
Cut lifting lug	1.00	0.1	EA	0	EA	0		
Instrumentation	14.00	0.1	EA	0	EA	0		
Remove blank plate	128.00	0.0	EA	0	EA	0		
Install stiffener plate	40.00	0.0	EA	0	EA	0		
10″ s80 BW	2.00	0.2	EA	0	EA	0		
10″ s80 fillet weld	4.00	0.2	EA	0	EA	0		

2.27.4 Street Steam Duct Sheet 4

Description	Historical			Estimate				
	MH	Quantity	Unit	Quantity	Unit	BM	IW	MW
Field Assemble Street Steam Duct						**0**	**0**	**0**
Handle 20'-6" diameter duct	10.00	1.3	EA	0	EA	0		
Weld duct; 20'-6" diameter×0.75" WT BW	1.15	8.3	LF	0	LF	0		
Handle drain pot	10.00	0.0	EA	0	EA	0		
Fit up and weld 0.75" BW	1.15	1.6	LF	0	LF	0		
Reinforcement pad	1.15	3.2	LF	0	LF	0		
Install and align embedded plates	8.00	0.1	EA	0	EA	0		
Weld together saddles and weld stiffeners	4.00	1.3	EA	0	EA	0		
Install saddles	4.00	0.2	EA	0	EA	0		
Bolt saddles to foundation	2.00	0.3	EA	0	EA	0		
Erect platform	10.00	0.7	EA	0	EA	0		
Place duct sections	10.00	0.3	EA	0	EA	0		
Fit up and weld 0.75" BW	1.15	43.1	LF	0	LF	0		
Grout support saddles	8.00	0.1	EA	0	EA	0		
Field weld w/slip ring 0.75" BW	1.15	3.3	LF	0	LF	0		
TED-06, 07, 08, 09 handle duct	10.00	1.6	EA	0	EA	0		
Fit up and weld 0.75" BW	1.15	9.6	EA	0	EA	0		
Install and align embedded plates	8.00	0.1	EA	0	EA	0		
Weld together saddles and weld stiffeners	4.00	1.3	EA	0	EA	0		
Install saddles	4.00	0.3	EA	0	EA	0		
Bolt saddles to foundation	2.00	0.3	EA	0	EA	0		
Place duct sections	10.00	0.4	EA	0	EA	0		
Fit up and weld 0.75" BW	1.15	40.0	LF	0	LF	0		
Field weld w/slip ring 0.75" BW	1.15	5.9	LF	0	LF	0		
Grout support saddles	8.00	0.1	EA	0	EA	0		
Place duct sections TED 05, 04	10.00	0.8	EA	0	EA	0		
Fit up and weld 0.75" BW	1.15	4.8	LF	0	LF	0		
Install and align embedded plates	8.00	0.1	EA	0	EA	0		
Weld together saddles and weld stiffeners	4.00	0.7	EA	0	EA	0		
Install saddles	4.00	0.2	EA	0	EA	0		
Bolt saddles to foundation	2.00	0.2	EA	0	EA	0		
Place duct	10.00	0.2	EA	0	EA	0		
Fit up and weld 0.75" BW	1.15	18.3	LF	0	LF	0		
Field weld w/slip ring 0.75" BW	1.15	2.2	LF	0	LF	0		
Place duct sections TED 03, 02	10.00	0.8	EA	0	EA	0		
Fit up and weld 0.75" BW	1.15	4.8	LF	0	LF	0		
Install and align embedded plates	8.00	0.1	EA	0	EA	0		
Weld together saddles and weld stiffeners	4.00	0.7	EA	0	EA	0		
Install saddles	4.00	0.2	EA	0	EA	0		
Bolt saddles to foundation	2.00	0.2	EA	0	EA	0		
Place duct	10.00	0.2	EA	0	EA	0		
Fit up and weld 0.75" BW	1.15	13.0	LF	0	LF	0		
Field weld w/slip ring 0.75" BW	1.15	1.5	LF	0	LF	0		
Grout support saddles	8.00	0.1	EA	0	EA	0		

2.27.5   Erect and Weld Turbine Exhaust Duct Risers Sheet 5

Description	Historical			Estimate				
	MH	Quantity	Unit	Quantity	Unit	BM	IW	MW
Erect and Weld Turbine Exhaust Duct Risers						**0**	**0**	**0**
Erect and Weld out Risers						**0**	**0**	**0**
Erect duct section Fit up and weld 0.75″ BW	10.00	0.14	EA	0	EA	0		
FFW	1.15	7.50	LF	0	LF	0		
FW	1.15	7.50	LF	0	LF	0		
Erect duct w/elbow Fit up and weld 0.75″ BW	10.00	0.14	EA	0	EA	0		
FFW	1.15	7.50	LF	0	LF	0		
FW	1.15	7.50	LF	0	LF	0		
Handle duct section w/elbow	10.00	0.03	EA	0	EA	0		
Fit up and weld 0.75″ BW	1.15	1.50	LF	0	LF	0		
Install and align embedded plates	8.00	0.11	EA	0	EA	0		
Weld together saddles and weld stiffeners	4.00	1.33	EA	0	EA	0		
Install saddles	4.00	0.33	EA	0	EA	0		
Bolt saddles to foundation	2.00	0.11	EA	0	EA	0		
Handle duct section	10.00	0.08	EA	0	EA	0		
Handle duct section w/elbow	10.00	0.03	EA	0	EA	0		
Fit up and weld 0.75″ BW								
FFW	1.15	1.50	LF	0	LF	0		
FW	1.15	1.50	LF	0	LF	0		

2.27.6 Risers, Tanks Sheet 6

Description	MH	Historical Quantity	Unit	Estimate Quantity	Unit	BM	IW	MW
Field Assemble Risers						0	0	0
Duct diameter 8'-6"								
Handle TED duct	10.00	0.14	EA	0	EA	0		
Fit up and weld 0.75" BW	1.15	6.94	LF	0	LF	0		
Handle 7051RD2	10.00	0.14	EA	0	EA	0		
Handle 7052RD2	10.00	0.14	EA	0	EA	0		
Fit up and weld 0.75" BW	1.15	7.50	LF	0	LF	0		
Expansion joint 8'-6"	10.00	0.28	EA	0	EA	0		
Handle 7021RD1	10.00	0.14	EA	0	EA	0		
Handle 7011TE2 and 7011EV2	10.00	0.28	EA	0	EA	0		
6'-10"								
Handle 7011TE1 and 7011EV1	10.00	0.28	EA	0	EA	0		
6'-10"								
Fit up and weld 0.75" BW	1.15	7.50	LF	0	LF	0		
Handle TED 1	10.00	0.03	EA	0	EA	0		
Expansion joint 8'6"	10.00	0.03	EA	0	EA	0		
Fit up and weld 0.75" BW	1.15	3.00	LF	0	LF	0		
Handle TED 1 6'-10"	10.00	0.03	EA	0	EA	0		
Handle 7046RD1	10.00	0.03	EA	0	EA	0		
Handle 7026RD2	10.00	0.03	EA	0	EA	0		
Fit up and weld 0.75" BW	1.15	1.50	LF	0	LF	0		
Expansion joint 8'-6"	10.00	0.06	EA	0	EA	0		
7026RD1 Handle 6'-10"	10.00	0.03	EA	0	EA	0		
7016TE2 and 7016TE 1 6'-10"	10.00	0.06	EA	0	EA	0		
7016EV1 and 7016EV2 6'-10"	10.00	0.06	EA	0	EA	0		
Fit up and weld 0.75" BW	1.15	7.50	LF	0	LF	0		
Set and Assemble Shop Tanks						0	0	0
Install and align embedded plates	8.00	0.11	EA	0	EA	0		
Weld together saddles and weld	4.00	0.33	EA	0	EA	0		
stiffeners								
Install saddles	4.00	0.06	EA	0	EA	0		
Bolt	2.00	0.06	EA	0	EA	0		
saddles to foundation								
Base plates	2.00	0.06	EA	0	EA	0		
Set condensate tank	40.00	0.03	EA	0	EA	0		
Set deaerator tank								
3'-11" Diameter	20.00	0.03	EA	0	EA	0		
8'-6" Diameter	20.00	0.03	EA	0	EA	0		
Fit up and weld								
0.5" × 25	0.75	0.69	LF	0	LF	0		
0.5" × 54	0.75	1.50	LF	0	LF	0		
Perforated plate	20.00	0.06	EA	0	EA	0		
Wire mesh	20.00	0.06	EA	0	EA	0		

2.27.7 Pumps, Expansion Joint, Test Sheet

Description	Historical			Estimate				
	MH	Quantity	Unit	Quantity	Unit	BM	IW	MW
Pumps						0	0	0
Set pump skid and pumps	120.0	1.00	EA	0	EA	0		
Install Expansion Joint at Turbine						0	0	0
Weld together upper dog bone holder								
Handle holder plates	4.0	0.56	EA	0	EA	0		
Handle corner section	4.0	0.22	EA	0	EA	0		
Fit up and weld holder plates 1.8″ thick	4.0	0.61	EA	0	EA	0		
Weld corner plates 1.1/16″ thick and 7/8″ thick	4.2	0.49	LF	0	LF	0		
Internal braces 5″ diameter std pipe	4.0	0.61	EA	0	EA	0		
Fillet weld	2.0	1.33	EA	0	EA	0		
Set upper dog bone to steam turbine duct	40.0	0.06	EA	0	EA	0		
Fit up and weld to SPX connection 3/4″ thick	1.1	4.77	LF	0	LF	0		
Weld SPX to turbine exhaust duct 2 1/4″ thick	2.5	4.77	LF	0	LF	0		
Install dog bone templates	2.0	0.22	EA	0	EA	0		
Rod bar	2.0	0.44	EA	0	EA	0		
Remove steel dog bone templates	2.0	0.22	EA	0	EA	0		
Install rubber at dog bone	1.0	4.77	EA	0	EA	0		
Clamping piece splice	1.0	2.96	EA	0	EA	0		
Field erection bolts	0.2	11.56	EA	0	EA	0		
Pneumatic Testing						0	0	0
Pneumatic testing	1200	0.03	TEST	0	TEST	0		

2.28 Air Cooled Condenser-Installation Man Hours

Facility-Combined Cycle Power Plant	Historical			Estimate	Actual			Actual		
Description	BM	IW	PF	MH/ Module	MH/ Module	BM	IW	PF		
Erect structural steel modules 1-6	0	0	0	0.0	263.1		9473			
Fan duct support steel modules 1-6	0	0	0	0.0	4.0		143			
Preassemble stairways at grade and set and secure	0	0	0	0.0	11.4		410			
Preassemble external walkways at grade	0	0	0	0.0	18.6		671			
Upper structure A-frame steelwork	0	0	0	0.0	294.7		10608			
Fan inlet cowl and fan screens	0	0	0	0.0	92.4		3326			
Motor support bridge	0	0	0	0.0	85.6		3082			
Erect and weld tube bundles	0	0	0	0.0	163.1	5872				
Erect and weld condensate headers	0	0	0	0.0	203.9	7340				
Erect and weld steam distribution manifolds	0	0	0	0.0	192.8			6942		
Erect and weld IC equipment pipe	0	0	0	0.0	81.7			2940		
Erect and weld IC street pipe	0	0	0	0.0	155.7			5606		
Erection and sheeting of wind walls	0	0	0	0.0	173.0		6230			
Erect and weld turbine exhaust duct	0	0	0	0.0	163.3	5877				
Field assemble street steam duct	0	0	0	0.0	272.4	9805				
Erection and welding of turbine exhaust duct risers	0	0	0	0.0	107.9	3884				
Set and assemble shop tanks	0	0	0	0.0	8.8	315				
Pumps	0	0	0	0.0	120.0	4320				
Install expansion joint at turbine	0	0	0	0.0	43.3	1560				
Pneumatic testing	0	0	0	0.0	33.3	1200				
Equipment/vendor piping installation man hours	**0**	**0**	**0**	0.0	**2489.0**	**40,173**	**33,943**	**15,489**		
(# of streets X modules)	**0**	**0**	**0**	0	**2489.0**					
(MH/module X # of streets X modules)=MH for ACC	0	0	0	0	89,605	40,173	33,943	15,489		

2.29 General Scope of Field Work Required for Each Surface Condenser

Surface Condenser
  Scope of Work-Field Erection

Low friction bearing plates
Off load components
Condenser assembly
  Assemble half shell
  Dome to condenser shell
  Condensate outlet sumps
  BW and thermal sleeves
  Anchor bolt seal rings
  Install water boxes
  Bolt condenser to foundation
Install dog bone expansion joint
Roll and jack condenser in Place

2.30 Surface Condenser Estimate Data

2.30.1 Bearing Plates, Off Load Components Sheet 1

Description	MH	Unit
Low Friction Bearing Plates		
Shim plates	10.00	EA
Set plates	10.00	EA
Off Load Components		
Left/right half shells, half dome, and water boxes	6.20	ton
Condensate sumps, expansion joints, and miscellaneous	20.00	EA

2.30.2   Condenser Assembly/Erection Sheet 2

Description	MH	Unit
Condenser Assembly		
Assemble Half Shell		
Align/bolt up guide clips	0.50	EA
Fit up and weld 5/8″ groove—angles, seams, and plates	1.43	LF
Remove temporary bracing	20.00	Lot
Dome to Condenser Shell		
Rig and set dome	60.00	EA
Fit up and weld dome sides to landing bar—groove weld	1.55	LF
Fit up and weld dome sides to landing bar—groove weld	1.44	LF
Fit up/prep and weld grid stiffeners to baffle landing bar—fillet weld	0.68	LF
Condensate Outlet Sumps		
Cut holes for sumps	8.80	EA
Rig and place sumps	30.00	EA
Fit up/prep and weld sump nozzle to bottom shell—1/2″ fillet weld	0.98	LF
BW and Thermal Sleeves		
Install nozzles	8.00	EA
Fit up/prep and weld nozzle to shell—1/2″ fillet weld	2.00	LF
Anchor Bolt Seal Rings		
Place seal rings	4.00	EA
Weld seal rings—1/2″ fillet weld	2.25	LF
Install Water Boxes		
Rig and place boxes	20.00	EA
Install gaskets	8.00	EA
Bolt boxes in place	0.29	EA
Bolt Condenser to Foundation		
Anchor bolts-3″	3.3	EA
Weld nuts	1.0	EA
Grout supports	9.0	EA

2.30.3   Steam Jet Air Injectors Sheet 3

Description	MH	Unit
Install Condenser Steam Jet Air Injectors		
Ejector, silencer, and support	20.0	EA
4″ 300# Valve	2.6	EA
14″ 150# Valve	9.1	EA
2″ Ejector	1.9	EA
6″ Ejector	5.2	EA
2″ 300# Valve	1.3	EA
6″ 150# Valve	5.1	EA
Piping	170.1	Lot

2.30.4 Expansion Joint, Place Condenser Sheet 4

Description	MH	Unit
Install Dog Bone Expansion Joint		
Handle, fit up/prep and weld sections together—3/4″ groove weld	60.0	EA
Rig and place exp joint	60.0	EA
Fit up/prep and weld expansion joint to dome	2.0	LF
Install rubber belt	1.0	LF
Install liner	0.5	LF
Install keepers and bolt	0.5	EA
Roll and Jack Condenser in Place		
Timber cribbing to support condenser		
12″×12″×8″ Timber	4.0	EA
8″×8″×8″ Timber	4.0	EA
300 ton Hillman rollers	8.0	EA
Channel support		
W18×97×54′	9.0	EA
C10×30×54′	9.0	EA
Jack unit in place	52.0	Unit

2.31 Surface Condenser Installation Estimate

2.31.1 Bearing Plates, Off Load Components Sheet 1

Description	MH	Historical Quantity	Unit	Estimate Quantity	Unit	BM
Low Friction Bearing Plates						0
Shim plates	10.00	6	EA	0	EA	0
Set plates	10.00	6	EA	0	EA	0
Off Load Components						0
Left/right half shells	27.50	2	ton	0	ton	0
Half dome	27.50	2	ton	0	ton	0
Inlet/outlet water boxes	27.50	2	ton	0	ton	0
Return water boxes	27.50	2	ton	0	ton	0
Condensate outlet sumps	20.00	2	EA	0	EA	0
Expansion joints	20.00	1	EA	0	EA	0
Miscellaneous	20.00	1	EA	0	EA	0

2.31.2 Condenser Assembly/Erection Sheet 2

Description	Historical			Estimate		
	MH	Quantity	Unit	Quantity	Unit	BM
Condenser Assembly/Erection						**0**
Assemble Half Shell						
Rig and set half shell-left/right	60.00	2	EA	0	EA	0
Align/bolt up guide clips	0.50	144	EA	0	EA	0
Fit up and weld bottom seam 5/8″ groove	1.43	40	LF	0	LF	0
Fit up and weld bottom angles	1.43	80	LF	0	LF	0
Fit up/prep and weld end seams—5/8″ groove	1.43	32	LF	0	LF	0
Fit up/prep and weld bottom support plates—5/8″ groove	1.43	32	LF	0	LF	0
Remove temporary bracing	20.00	1	Lot	0	Lot	0
Rig and set dome shell-left/right	20.00	2	EA	0	EA	0
Align/bolt up guide clips	0.50	32	EA	0	EA	0
Fit up/prep and weld vertical seams—5/8″ groove	1.43	16	LF	0	LF	0
Fit up and weld top/bottom grid stiffeners—5/8″ groove	1.43	48	LF	0	LF	0
Remove temporary bracing	20.00	1	Lot	0	Lot	0
Dome to Condenser Shell						
Rig and set dome	60.00	1	EA	0	EA	0
Align/bolt up guide clips	0.50	216	EA	0	EA	0
Fit up and weld dome sides to landing bar—groove weld	1.55	80	LF	0	LF	0
	1.44	48	LF	0	LF	0
Fit up/prep and weld grid stiffeners to baffle landing bar	0.68	28	LF	0	LF	0
Remove temporary bracing	20.00	1	Lot	0	Lot	0
Condensate Outlet Sumps						
Cut holes for sumps	8.80	2	EA	0	EA	0
Rig and place sumps	30.00	2	EA	0	EA	0
Fit up/prep and weld sump nozzle to bottom shell	0.98	64	LF	0	LF	0
BW and Thermal Sleeves						
Install nozzles	8.00	12	EA	0	EA	0
Fit up/prep and weld nozzle to shell	2.00	12	LF	0	LF	0
Anchor Bolt Seal Rings						
Place seal rings	4.00	16	EA	0	EA	0
Weld seal rings	2.25	16	LF	0	LF	0
Install Water Boxes						
Rig and place boxes	20.00	2	EA	0	EA	0
Install gaskets	8.00	16	EA	0	EA	0
Bolt boxes in place	0.29	248	EA	0	EA	0
Bolt Condenser to Foundation						
Anchor bolts-3″	3.3	12	EA	0	EA	0
Weld nuts	1.0	12	EA	0	EA	0
Grout supports	9.0	12	EA	0	EA	0

2.31.3   Steam Jet Air Injectors Sheet 3

	Historical			Estimate		
Description	MH	Qty	Unit	Qty	Unit	BM
Install Condenser Steam Jet Air Injectors						**0**
14″ hogging ejector						
Silencer support	20.0	1	EA	0	EA	0
Silencer	20.0	1	EA	0	EA	0
Ejector	20.0	1	EA	0	EA	0
4″ 300# Valve	2.6	1	EA	0	EA	0
14″ 150# Valve	9.1	1	EA	0	EA	0
2″ Ejector	1.9	2	EA	0	EA	0
6″ Ejector	5.2	2	EA	0	EA	0
2″ 300# Vvalve	1.3	2	EA	0	EA	0
6″ 150# Valve	5.1	4	EA	0	EA	0
Piping	170.1	1	Lot	0	Lot	0

2.31.4   Expansion Joint, Place Condenser Sheet 4

	Historical			Estimate		
Description	MH	Quantity	Unit	Quantity	Unit	BM
Install Dog Bone Expansion Joint						**0**
Handle, fit up/prep and weld sections—3/4″ groove weld	60.0	2	EA	0	EA	0
Rig and place expansion joint	60.0	1	EA	0	EA	0
Fit up/prep and weld expansion joint to dome	2.0	100	LF	0	LF	0
Install rubber belt	1.0	100	LF	0	LF	0
Install liner	0.5	100	LF	0	LF	0
Install keepers and bolt	0.5	100	EA	0	EA	0
Roll and Jack Condenser in Place						**0**
Timber cribbing to support condenser						
12″×12″×8″ Timber	4.0	28	EA	0	EA	0
8″×8″×8″ Timber	4.0	28	EA	0	EA	0
300 ton Hillman rollers	8.0	4	EA	0	EA	0
Channel support						
W18×97×54′	9.0	4	EA	0	EA	0
C10×30×54′	9.0	4	EA	0	EA	0
Jack unit in place	52.0	1	Unit	0	Unit	0

2.32   Surface Condenser-Equipment Installation Man Hours

Facility-Combined Cycle Power Plant	Actual	Estimated
	MH	BM
Low friction bearing plates	120	0
Off load components	300	0
Condenser assembly/erection	1801	0
Install condenser steam jet air injectors	279	0
Install dog bone expansion joint	580	0
Roll and jack condenser in place	380	0
Equipment/vendor piping installation man hours	3460	0

2.33   Structural Steel Man Hour Table

Facility Class—Combined Cycle Power Plant
   Structural Steel and Miscellaneous Iron

Erect Structural Steel	MH/ton
Erect Structural Steel; <=20 tons	
Light—0–19 lb/ft	28.0
Medium—20–39 lb/ft	24.0
Heavy—40–79 lb/ft	20.0
X heavy—80–120 lb/ft	16.0
Erect Structural Steel; 20 tons > tons <= 100 tons	
Light—0–19 lb/ft	21.0
Medium—20–39 lb/ft	18.0
Heavy—40–79 lb/ft	15.0
X heavy—80–120 lb/ft	12.0
Erect Structural Steel; >100 tons	
Light—0–19 lb/ft	16.8
Medium—20–39 lb/ft	14.4
Heavy—40–79 lb/ft	13.2
X heavy—80–120 lb/ft	11.8
	MH/SF
Platform framing	0.15
	MH/LF
Handrail and toe plate	0.25
	MH/SF
Floor grating	0.20
	MH/LF
Stair treads	0.85
Ladders	**MH/LF**
Straight ladder	0.30
Caged ladders	0.35

Simple Cycle Power Plant Equipment

3.1 LM 6000 Gas Turbine Generator Set and Selective Catalytic Reduction

The LM 6000 is designed for simple-cycle and cogeneration installations. It can start and stop easily for "peaking" and "dispatched" applications. Additionally, quick dispatch ability is available in simple cycle applications with the 10-min fast start feature.

The gas turbine is a two-shaft gas turbine engine equipped with low-pressure compressor, high-pressure compressor, combustor, high-pressure turbine, and low-pressure turbine.

The compressor compresses combustion air to the combustor where the fuel is mixed with the combustion air and burned. Hot exhaust gases then enter the power turbine where the gases expand across the turbine blades, rotating a shaft to power the electric generator. After exiting the combustion turbine, the hot exhaust gases are then sent through the selective catalytic reduction (SCR) unit to reduce oxides of nitrogen in the exhaust and an oxidation catalyst to reduce organic compounds and carbon monoxide in the exhaust.

A nitrogen-based reagent such as ammonia or urea is injected into the ductwork, downstream of the combustion unit. The waste gas mixes with the reagent and enters a reactor module containing catalyst. The hot flue gas and reagent diffuse through the catalyst. The reagent reacts selectively with the Nox within a specific temperature range and in the presence of the catalyst and oxygen.

Temperature, the amount of reducing agent, injection grid design, and catalyst activity are the main factors that determine the actual removal efficiency.

Industrial Piping and Equipment Estimating Manual. http://dx.doi.org/10.1016/B978-0-12-813946-2.00003-4

3.2 General Scope of Field Work Required for LM 6000 Gas Turbine Generator

LM 6000 Turbine Generator
Scope of Work-Field Erection

LM 6000 gas turbine—generator set
Scope of work-field erection
Receive and stage equipment
Check and prepare main skid foundation
Set and level turbine skid
Set and level generator skid
Install gating
Set and level auxiliary skid
Install/grout generator into package
Prep/install roof skid
Install VBV hood and silencer
Install generator exhaust ventilation, and air/oil separator
Install air filter support structure
Install air filter assembly
Install ventilation recirculate antiicing
Install air filter ladders and platforms
Prep air filter for filtration media installation
Install filtration media
Set level and install subskids
Turbine generator alignment
Final turbine enclosure dress out
Final generator enclosure dress out
Final auxiliary skid dress out
Pour grout-grout main and auxiliary skids including subskids
Install line side and neutral cubicle

3.3   LM 6000 Turbine Generator Estimate Data

3.3.1   Place Skids, Install Generator, and Roof Skid Sheet 1

Description	MH	Unit
Main, Turbine, Generator and Auxiliary Skids		
Place and level shims—main, turbine, generator, and auxiliary skids	1.75	EA
Install grout dams—main skid	2.00	EA
Set and level turbine, generator, auxiliary skid and generator	2.50	ton
Install grating between main and auxiliary skid and floor grating	0.20	SF
Install/Grout Generator into Package		
Roof/end wall removal/installation	10.00	EA
Bolt sole plates/install; hold down bolts	2.00	EA
Grout generator	1.40	SF
Prep/Install Roof Skid		
Install ventilation fan silencers to vent fans	24.00	ton
Install ventilation stack caps to air silencers and VBV elbow	50.00	ton
Install generator ventilation back draft dampers and set roof skid	80.00	EA
Install VBV hood and silencer	50.00	ton
Receive and stage equipment	400	UNIT

3.3.2   Generator Exhaust Ventilation, Air Filter, Antiicing, Media, and Platforms Sheet 2

Description	MH	Unit
Install Generator Exhaust Ventilation, and Air/Oil Separator		
Install damper, silencer, hood and ventilation assembly, and separator	20.00	EA
Torque bolts/seal joints	8.00	Joints
Install Air Filter Assembly		
Install left/right hand air filter module, plenum, and transition	6.40	ton
Install air filter support structure	18.95	ton
Install Ventilation Antiicing		
Install ventilation gravity close dampers and elbow/transition duct	10.00	EA
Install antiicing plenum, exhaust damper, and antiicing rake and plenum	20.00	EA
Install Media and Ladders and Platforms		
Install platforms; grating and handrails	20	ton
Install ladders; fan, filter house	0.35	LF
Prep air filter for filtration media installation	80	JOB
Install filtration media	80	UNIT

3.3.3   Subskids, Generator Alignment, Final Enclosure, Generator, and Skids Dress Out Sheet 3

Description	MH	Unit
Set level and install subskids	120	SKID
Turbine generator alignment	2.64	ton
Final Turbine Enclosure, Generator, and Auxiliary Skid Dress Out		
Install the stage and CDP piping (DLE only)	60	UNIT
Install coupling/guard	20	EA
Install expansion joints-VBV duct, inlet volute, exhaust	20	EA
Install ductwork	80	JOB
Connect LO supply/return and CW supply/return to generator	80	UNIT
Battery cabinet/batteries	10	EA
Connect LO, liquid fuel, gas fuel and water supply/ return	120	UNIT
Pour grout-grout main and auxiliary skids including subskids	480	UNIT
Install line side and neutral cubicle	60	UNIT

3.4   LM 6000 Turbine Generator Equipment Installation Estimate

3.4.1   Place Skids, Install Generator, and Roof Skid Sheet 1

Estimate—Receive Equipment		Historical		Estimate				
Description	MH	Quantity	Unit	Quantity	Unit	BM	IW	MW
Receive and stage equipment	400	1	JOB	0	JOB	0	0	
Estimate—Place Skids		**Historical**		**Estimate**				
Main, turbine, generator, and auxiliary skids						0	0	0
Place-main, turbine, generator and auxiliary skids	1.75	16	EA	0	EA			0
Install grout dams—main skid	2.00	16	EA	0	EA			0
Set and level combustion turbine skid	2.50	56	ton	0	ton		0	
Set and level generator skid	2.50	48	ton	0	ton		0	
Install grating between main and auxiliary skid	0.20	200	SF	0	SF		0	
Set and level auxiliary skid	2.50	32	ton	0	ton		0	
Estimate—Install Generator		**Historical**		**Estimate**				
Install/Grout Generator into Package						0	0	0
Roof/end wall removal/ installation	10.00	2	EA	0	EA		0	
Set generator into package	2.50	48	ton	0	ton		0	
Install floor grating	0.20	140	SF	0	SF		0	
Bolt sole plates/install; hold down bolts	2.00	16	EA	0	EA		0	
Grout generator	1.40	100	SF	0	SF			0
Estimate - Install Roof Skid		**Historical**		**Estimate**				
Prep/install roof skid						0	0	0
Install ventilation fan silencers to vent fans	24.00	2	ton	0	ton	0		
Install ventilation stack caps to air silencers	50.00	1	ton	0	ton	0		
Install VBV transition elbow	50.00	1	ton	0	ton	0		
Install generator dampers and set roof skid	80.00	1	EA	0	EA	0		
Install VBV hood and silencer	50.00	3	ton	0	ton	0		

3.4.2   Generator Exhaust Ventilation, Air Filter, Antiicing, Media, and Platforms Sheet 2

Estimate—Generator Exhaust Ventilation		Historical			Estimate				
Description	MH	Quantity	Unit	Quantity	Unit	BM	IW	MW
Install generator exhaust ventilation						0	0	0
Install exhaust damper to transition pecs	20.00	1	EA	0	EA			
Install generator exhaust silencer	20.00	1	EA	0	EA	0		
Install exhaust hood to silencer assembly	20.00	1	EA	0	EA	0		
Install exhaust ventilation assembly	20.00	1	EA	0	EA	0		
Torque bolts/seal joints	8.00	5	Joints	0	Joints	0		
Install tank air/oil separator to enclosure	20.00	1	EA	0	EA	0		
Estimate—Air Filter Assembly		**Historical**			**Estimate**			
Install air filter assembly						0	0	0
Install air filter plenum on main unit	6.40	13	ton	0	ton	0		
Install left hand air filter module	6.40	19	ton	0	ton	0		
Install air inlet silencer/transition	6.40	6	ton	0	ton	0		
Install right hand air filter module	6.40	9	ton	0	ton	0		
Install air filter support structure	18.95	10	ton	0	ton	0		
Estimate—Antiicing		**Historical**			**Estimate**			
Install ventilation antiicing						0	0	0
Install ventilation gravity close dampers (2)	10.00	2	EA	0	EA	0		
Install ventilation antiicing plenum	20.00	1	EA	0	EA	0		
Install common exhaust damper	20.00	1	EA	0	EA	0		
Install motor operated dampers (2)	20.00	2	EA	0	EA	0		
Install R/H and L/H antiicing rake	20.00	2	EA	0	EA	0		
Install R/H and L/H antiicing plenum	20.00	2	EA	0	EA	0		
Install elbow and transition duct	10.00	4	EA	0	EA	0		
Estimate-Media and Platforms		**Historical**			**Estimate**			
Install media and ladders and platforms						0	0	0
Install platforms; grating and handrails	20.00	10	ton	0	ton	0		
Install ladders; fan, filter house	0.35	75	LF	0	LF	0		
Prep air filter for filtration media installation	80.00	1	JOB	0	JOB	0		
Install filtration media	80.00	1	UNIT	0	UNIT	0		

3.4.3 Subskids, Generator Alignment, Final Enclosure, Generator, and Skids Dress Out Sheet 3

	Historical			Estimate				
Description	MH	Quantity	Unit	Quantity	Unit	BM	IW	MW
Set level and install						**0**	**0**	**0**
Subskids								
Liquid fuel duplex filter skid	120.00	1	SKID	0	SKID		0	
Gas fuel filter skid	120.00	1	SKID	0	SKID		0	
Steam skid	120.00	1	SKID	0	SKID		0	
Fin-fan lube oil coolers skid	120.00	1	SKID	0	SKID		0	
Calorimeter skid	120.00	1	SKID	0	SKID		0	
Turbine generator alignment	2.64	121	ton	0	ton			**0**
Final turbine enclosure Dress out						**0**	**0**	**0**
Install the stage and CDP piping (DLE only)	60.00	1	UNIT	0	UNIT	0		
Install coupling/ guard	20.00	1	EA	0	EA			0
Install expansion joints	20.00	3	EA	0	EA	0		
Install floor grating	0.20	100	SF	0	SF	0		
Final generator enclosure dress out						**0**	**0**	**0**
Install coupling/ guard	20.00	1	EA	0	EA			0
Install ductwork	80.00	1	JOB	0	JOB	8		
Floor grating	0.20	200	SF	0	SF	0		
LO supply/return and CW supply/return	80.00	1	UNIT	0	UNIT	0		
Final auxiliary skid dress out						**0**	**0**	**0**
Install coupling/ guard	20.00	1	EA	0	EA	0		0
Install ductwork	80.00	1	JOB	0	JOB	0		
Floor grating	0.20	200	SF	0	SF	0		
Battery cabinet/ batteries	10.00	2	EA	0	EA	0		
LO, liquid fuel, gas fuel, and water supply/return	120.00	1	UNIT	0	UNIT	0		
Grout main and auxiliary skids including subskids	480.00	1	UNIT	0	UNIT			**0**
Install line side and neutral cubicle	60.00	1	UNIT	0	UNIT		**0**	

3.5 LM 6000 Turbine Generator-Equipment Installation Man Hours

	Actual	Estimated			
Facility-Simple Cycle Power Plant	MH	BM	IW	MW	MH
Receive and stage equipment	400	0	0		0
Main, turbine, generator, and auxiliary skids	440	0	0	0	0
Install/grout generator into package	340	0	0	0	0
Prep/install roof skid	340	0	0	0	0
Install generator exhaust ventilation	140	0	0	0	0
Install air filter assembly	480	0	0	0	0
Install ventilation antiicing	220	0	0	0	0
Install media, ladders, and platforms	380	0	0	0	0
Set level and install subskids	600	0	0	0	0
Turbine generator alignment	320			0	0
Final turbine enclosure dress out	160	0	0	0	0
Final generator enclosure dress out	220	0	0	0	0
Final auxiliary skid dress out	820	0	0	0	0
Equipment installation man hours	**4860**	**0**	**0**	**0**	**0**

3.6 General Scope of Field Work Required for Each Selective Catalytic Reduction

Selective catalytic reduction
Scope of work-field erection

Receive and stage equipment
Sole plates/AB
Stair tower
Module/duct
Ammonia grid
CO Catalyst
SCR seal frame
Ladders and platforms
Liner plates
AIG supply pipe
Tempering air fans
SCR catalyst module
Install catalyst
Ammonia vaporizer
Electric air heater
Aqueous ammonia skid
Ammonia vaporization skid
Stack
Exhaust stack silencer
Acoustical shroud

3.7 Selective Catalytic Reduction Estimate Data

3.7.1 Receive Equipment, Modules, Ducts, Platforms, and AIG Pipe Sheet 1

Description	MH	Unit
Receive and stage equipment	440.00	UNIT
Shim and set sole plates	4.00	EA
Anchor bolts	0.65	EA
Erect Modules, Ducts & and Expansion Joints		
Install duct and expansion joint	60.00	EA
Field joint	0.40	LF
Install module	6.40	ton
Ammonia Grid		
Ammonia grid and CO catalyst module and SCR frame	20.00	ton
Remove shipping angles	40.00	MODULE
Install tie plate	40.00	MODULE
Liner Plates		
Field liner plates	10.00	EA
Ladders and Platforms		
Platform	0.15	SF
Handrail	0.25	LF
Ladders	0.35	LF
AIG Supply Pipe		
10" s10s TP304 pipe	2.40	LF
10" s10s TP304 fit up/BW	8.80	EA
10" 150# bolt up	4.50	EA
10" pipe support	8.00	EA

3.7.2 Tempering Air Fans, Vaporizer, Heater, Skids, and Stack/ Silencer Sheet 2

Description	MH	Unit
Tempering Air Fans		
Tempering air fans	30.00	EA
Tempering air silencer	20.00	EA
Tempering air filter w/rain hood	20.00	EA
Bolt connections	12.00	EA
Purge air fan	10.00	EA
Purge air filter w/rain hood	10.00	EA
Air duct	0.60	LF
Damper	10.00	EA
Bolt connections	10.00	EA
8" equalizing pipe	2.20	LF
6" equalizing pipe	1.50	LF
8" valve	3.60	EA
6" valve	2.70	EA
8" 150# bolt up	3.60	EA
6" 150# bolt up	2.70	EA
Ammonia vaporizer	40.00	EA
Hanger support	20.00	EA
Electric Air Heater		
Electric air heater	40.00	EA
Aqueous Ammonia and ammonia vaporization skids	80.00	SKID
Stack	420.00	EA
Exhaust Stack Silencer		
Support brackets	8.00	EA
Baffles	12.00	EA
Wall bracket	8.00	EA
Baffle guide	6.00	EA
Perforated liner	40.00	EA
Acoustical Shroud		
Panels	12.00	EA
Structural tubing	8.00	EA
Wall flashing	10.00	EA

3.7.3 Stair Tower, Catalyst Modules, Acoustical Shroud Sheet 3

Description	MH	Unit
Stair Tower		
Set structural	10	EA
Elevation X-brace	2	EA
Bolt connections	0.5	EA
Stairs	0.85	EA
Handrail	0.25	LF
Platform	0.15	SF
AB	0.5	EA
Catalyst Modules		
Catalyst modules	10	EA
Install catalyst	4	EA

3.8 Selective Catalytic Reduction Equipment Installation Estimate

3.8.1 Receive Equipment, Modules, Ducts, Platforms, and AIG Pipe Sheet 1

	Historical			Estimate		
Description	MH	Quantity	Unit	Quantity	Unit	BM
Receive and stage	440.00	1.0	UNIT	0.0	UNIT	**0**
equipment						
Shim, set sole plates,						**0**
and anchor bolts						
Shim and set sole plates	4.00	22.0	EA	0.0	EA	0
Anchor bolts	0.65	104.0	EA	0.0	EA	0
Erect Modules, ducts,						**0**
and expansion joints						
Expansion joint	60.00	1.0	EA	0.0	EA	0
4100 duct	60.00	1.0	EA	0.0	EA	0
4200 duct	60.00	1.0	EA	0.0	EA	0
Field joint	0.40	220.4	LF	0.0	LF	0
4200 module	6.40	9.4	ton	0.0	ton	0
4300 module	6.40	15.0	ton	0.0	ton	0
4400 module	6.40	26.5	ton	0.0	ton	0
4500 module	6.40	31.0	ton	0.0	ton	0
Field joint	0.40	800.0	LF	0.0	LF	0
Ammonia Grid						
Ammonia grid module	20.00	1.0	ton	0.0	ton	0
Field joint	0.40	204.0	LF	0.0	LF	0
Remove shipping angles	40.00	1.0	MODULE	0.0	MODULE	0
Install tie plate	40.00	1.0	MODULE	0.0	MODULE	0
CO Catalyst						
CO catalyst module	20.00	1.0	ton	0.0	ton	0
Field joint	0.40	204.0	LF	0.0	LF	0
SCR Seal Frame						
Seal frame	20.00	1.0	ton	0.0	ton	0
Field joint	0.40	204.0	LF	0.0	LF	0
Liner Plates						
Field liner plates	10.00	176.0	EA	0.0	EA	0
Ladders and platforms						**0**
Platform at EL 19′,	0.15	750.0	SF	0.0	SF	0
30′, 32′						
Handrail	0.25	240.0	LF	0.0	LF	0
Ladders	0.35	81.0	LF	0.0	LF	0
Platform at EL 51′, 72′	0.15	180.0	SF	0.0	SF	0
Handrail	0.25	56.0	LF	0.0	LF	0
Ladders	0.35	62.0	LF	0.0	LF	0
AIG supply pipe						**0**
10″ s10s TP304 pipe	2.40	42.0	LF	0.0	LF	0
10″ s10s TP304 fit up/	8.80	3.0	EA	0.0	EA	0
BW						
10″ 150# bolt up	4.50	2.0	EA	0.0	EA	0
10″ pipe support	8.00	1.0	EA	0.0	EA	0

3.8.2   Tempering Air Fans, Vaporizer, Heater, Skids, and Stack/Silencer Sheet 2

Description	Historical			Estimate		
	MH	Quantity	Unit	Quantity	Unit	BM
Tempering air fans						**0**
Tempering air fans	30.00	2.0	EA	0.0	EA	0
Tempering air silencer	20.00	2.0	EA	0.0	EA	0
Tempering air filter w/rain hood	20.00	2.0	EA	0.0	EA	0
Bolt connections	12.00	4.0	EA	0.0	EA	0
Purge air fan	10.00	2.0	EA	0.0	EA	0
Purge air filter w/rain hood	10.00	2.0	EA	0.0	EA	0
Air duct	0.60	100.0	LF	0.0	LF	0
Damper	10.00	4.0	EA	0.0	EA	0
Bolt connections	10.00	10.0	EA	0.0	EA	0
8″ equalizing pipe	2.20	30.0	LF	0.0	LF	0
6″ equalizing pipe	1.50	30.0	LF	0.0	LF	0
8″ valve	3.60	3.0	EA	0.0	EA	0
6″ valve	2.70	3.0	EA	0.0	EA	0
8″ 150# bolt up	3.60	6.0	EA	0.0	EA	0
6″ 150# bolt up	2.70	6.0	EA	0.0	EA	0
Vaporizer, Air heater, and Ammonia skid						**0**
Ammonia Vaporizer						
Ammonia vaporizer	40.00	1.0	EA	0.0	EA	0
Hanger support	20.00	4.0	EA	0.0	EA	0
Electric Air Heater						
Electric air heater	40.00	1.0	EA	0.0	EA	0
Hanger support	80.00	4.0	EA	0.0	EA	0
Aqueous ammonia skid	80.00	1.0	SKID	0.0	SKID	0
Ammonia vaporization skid	80.00	1.0	SKID	0.0	SKID	0
Stack and Silencer						**0**
Stack	420.00	1.0	EA	0.0	EA	0
Exhaust Stack Silencer						
Support brackets	8.00	4.0	EA	0.0	EA	0
Baffles	12.00	4.0	EA	0.0	EA	0
Wall bracket	8.00	8.0	EA	0.0	EA	0
Baffle guide	6.00	8.0	EA	0.0	EA	0
Perforated liner	40.00	2.0	EA	0.0	EA	0
Acoustical Shroud						
Panels	12.00	6.0	EA	0.0	EA	0
Structural tubing	8.00	13.0	EA	0.0	EA	0
Wall flashing	10.00	4.0	EA	0.0	EA	0

3.8.3 Stair Tower, Catalyst Modules, Acoustical Shroud Sheet 3

	Historical			Estimate		
Description	MH	Quantity	Unit	Quantity	Unit	BM
Acoustical shroud						**0**
Panels	12.00	6.0	EA	0.0	EA	0
Structural tubing	8.00	13.0	EA	0.0	EA	0
Wall flashing	10.00	4.0	EA	0.0	EA	0
Stair tower						**0**
North/south elevation set structural	10.00	6.0	EA	0.0	EA	0
North/south elevation X-brace	2.00	6.0	EA	0.0	EA	0
East/west elevation set structural	10.00	6.0	EA	0.0	EA	0
East/west elevation X-brace	2.00	7.0	EA	0.0	EA	0
Bolt connections	0.50	144.0	EA	0.0	EA	0
Stairs	0.85	47.0	EA	0.0	EA	0
Handrail	0.25	240.0	LF	0.0	LF	0
Platform	0.15	290.0	SF	0.0	SF	0
AB	0.50	8.0	EA	0.0	EA	0
Catalyst modules						**0**
Catalyst modules	10.00	8.0	EA	0.0	EA	0
Install catalyst	4.00	80.0	EA	0.0	EA	0

3.9 Selective Catalytic Reduction-Equipment Installation Man Hours

	Actual	Estimated
Facility-Simple Cycle Power Plant	MH	BM
Receive and stage equipment	440	0
Shim, set sole plates, and anchor bolts	156	0
Erect modules, ducts, and expansion joints	3257	0
Ladders and platforms	264	0
AIG supply pipe	144	0
Tempering air fans	596	0
Vaporizer, air heater, and ammonia skid	640	0
Stack and silencer	908	0
Acoustical shroud	216	0
Stair tower	365	0
Catalyst modules	400	0
Equipment installation man hours	**7386**	**0**

Chiller and Cooling Tower

3.10   Chiller and Cooling Tower Estimate Data

3.10.1   Chiller and Cooling Tower Sheet 1

Description	MH	Unit
Chiller and pump skid	3.77	ton
GT lube oil cooling water skid	60.00	SKID
GT LO CW skid AB	6.0	SKID
Chiller skid AB	24.00	EA
Column AB	4.0	EA
Walkway	0.15	SF
Ladder	0.35	LF
Handrail	0.25	LF
Stairs	0.80	EA
Cooling tower (40 hp)	340.00	EA
Cooling tower fan A/B (40 hp)	20.00	EA
Chiller circulating water pump A/B (75 hp)	60.00	EA
Cooling tower chemical storage totes	4.00	EA

3.11   Chiller and Cooling Tower Equipment Installation Estimate

3.11.1   Chiller and Cooling Tower Sheet 1

| Description | Historical | | | Estimate | | | | |
	MH	Quantity	Unit	Quantity	Unit	BM	IW	MW
Chiller						0	0	0
Chiller skid	3.77	85.0	ton	0	ton		0	
Pump skid	3.77	35.0	ton	0	ton		0	
GT lube oil cooling water skid	60.00	1.0	SKID	0	SKID		0	
GT LO CW skid AB	6.00	4.0	SKID	0	SKID		0	
Chiller skid AB	24.00	1.0	SKID	0	EA		0	
Column AB	4.00	6.0	EA	0	EA		0	
Walkway	0.15	180.0	SF	0	SF		0	
Ladder	0.35	25.0	LF	0	LF		0	
Handrail	0.25	120.0	LF	0	LF		0	
Stairs	0.80	20.0	EA	0	EA		0	
Cooling tower (40 hp)	340.00	1.0	EA	0	EA	0	0	0
Cooling tower fan A/B (40 hp)	20.00	2.0	EA	0	EA	0	0	0
Chiller circulating water pump A/B (75 hp)	60.00	1.0	EA	0	EA	0	0	0
Cooling tower chemical storage totes	4.00	4.0	EA	0	EA	0	0	0

3.12 Chiller and Cooling Tower-Equipment Installation Man Hours

Facility-Simple Cycle Power Plant	Actual	Estimated			
	MH	BM	IW	MW	MH
Chiller	666	0	0	0	**0**
Cooling tower (40 hp)	340	0	0	0	**0**
Cooling tower fan A/B (40 hp)	40	0	0	0	**0**
Chiller circulating water pump A/B (75 hp)	60	0	0	0	**0**
Cooling tower chemical storage totes	16	0	0	0	**0**
Equipment installation man hours	**1122**	**0**	**0**	**0**	**0**

3.13 Balance of Plant Equipment Estimate Data

3.13.1 Tanks, Filters, Compressors and Generators, Skids, Pumps Sheet 1

Description	MH	Unit
Aqueous Ammonia Unloading/Storage		
Ammonia forwarding pump A/B skid	40.00	SKID
Ammonia storage tank, liquid drain tank	10.4	ton
Air compressor/dryer skids	20.00	ton
Ammonia forwarding pump A/B skid	40.00	SKID
Ammonia unloading skid	40.00	SKID
Fuel gas coalescing filters	40.00	EA
Fuel gas final and wastewater filters	40.00	EA
Compressed air receiver	20.00	ton
Fuel gas compressor skid 800 hp	12.00	ton
CEMS	35.50	ton
Fire pump and sodium injection skids	40.00	SKID
Fuel gas drains tank	10.40	EA
Oil/water separator	40.00	EA
Process wastewater sump	40.00	EA
Pumps		
Ammonia pump A/B	20.00	EA
Demin water transfer pumps A/B (7.5 hp)	20.00	EA
Process wastewater sump pump 2 hp	20.00	EA
Oil/water sump pump	40.00	EA
Service water pumps	20.00	EA

3.14 Balance of Plant Equipment Installation Estimate

3.14.1 Tanks, Filters, Compressors, Skids, Pumps Sheet 1

Description	MH	Historical Quantity	Unit	Estimate Quantity	Unit	BM	IW	MW
Aqueous Ammonia Unloading/Storage								
Ammonia forwarding pump A/B skid	40	1	SKID	0	SKID		0	
Aqueous ammonia storage tank	10.4	6.7	ton	0	ton		0	
Ammonia unloading skid	40	1	SKID	0	SKID		0	
Ammonia pump A/B	20	2	EA	0	EA		0	0
Demin water transfer pumps A/B (7.5 hp)	20	2	EA	0	EA		0	0
Fuel gas liquids drain tank	10.4	4.3	ton	0	ton		0	
Fuel gas coalescing filters	40	2	EA	0	EA		0	
Fuel gas final filter	40	2	EA	0	EA		0	
Air compressor skid A/B	20	1	ton	0	ton		0	
Compressed air receiver	20	1	ton	0	ton		0	
Compressed air dryer skid; 50 hp	20	1	ton	0	ton		0	
Process wastewater sump pump 2 hp	20	1	EA	0	EA		0	0
Process wastewater sump	40	1	EA	0	EA		0	
Fuel gas compressor skid 800 hp	12	25	ton	0	ton		0	
Fuel gas drains tank	10.4	2	ton	0	ton		0	
Oil/water separator	40	1	EA	0	EA		0	
Oil/water sump pump	40	1	EA	0	EA		0	0
Wastewater filter	40	2	EA	0	EA		0	
Service water pumps	20	2	EA	0	EA		0	0
Sodium hypochlorite injection skid	40	1	EA	0	EA		0	
Fire pump skid	40	1	SKID	0	SKID		0	
CEMS	35.50	2.3	ton	0	ton	0		

3.15 Balance of Plant-Equipment Installation Man Hours

	Actual	Estimated			
Facility-Simple Cycle Power Plant	**MH**	**BM**	**IW**	**MW**	**MH**
Shop Fabricated Tanks					
Aqueous ammonia storage tank	70	0	0	0	**0**
Fuel gas liquids drain tank	45	0	0	0	**0**
Fuel gas drains tank	21	0	0	0	**0**
Filters					
Fuel gas coalescing filters	80	0	0	0	**0**
Fuel gas final filter	80	0	0	0	**0**
Wastewater filter	80	0	0	0	**0**
Compressors and Generators					
Fuel gas compressor skid 800 hp	300	0	0	0	**0**
Equipment and Skids					
Ammonia forwarding pump A/B skid	40	0	0	0	**0**
Ammonia unloading skid	40	0	0	0	**0**
Air compressor skid A/B	19	0	0	0	**0**
Compressed air receiver	20	0	0	0	**0**
Compressed air dryer skid; 50 hp	22	0	0	0	**0**
Sodium hypochlorite injection skid	40	0	0	0	**0**
Fire pump skid	40	0	0	0	**0**
CEMS	80	0	0	0	**0**
Process wastewater sump	40	0	0	0	**0**
Oil/water separator	40	0	0	0	**0**
Pumps and Drivers					
Ammonia pump A/B	40	0	0	0	**0**
Demin water transfer pumps A/B (7.5 hp)	40	0	0	0	**0**
Process wastewater sump pump 2 hp	20	0	0	0	**0**
Oil/water sump pump	40	0	0	0	**0**
Service water pumps	40	0	0	0	**0**
Equipment installation man hours	**1236**	**0**	**0**	**0**	**0**

Refinery and Hydrogen Plant Equipment

4.1 Steam Methane Reformer and Pressure Swing Adsorbers (PSAs)

Steam Methane Reformer

Feedstock

Natural gas is the most common feedstock in steam reformers; naphtha and refinery off-gas may also be used.

Main Component

Furnace

The furnace is where the process of liberating hydrogen from natural gas and steam begins. The gas and steam mixture travels down into reformer tubes that hang in vertical rows surrounded by gas burners that heat the mixture. The reformer tubes are full of nickel catalyst, which triggers a reaction causing the methane in natural gas to react with water vapor to form hydrogen, carbon monoxide, and carbon dioxide.

Pressure Swing Adsorbers (PSAs)

The PSAs are used to filter out remaining traces of carbon monoxide, carbon dioxide, and methane from the hydrogen. These leftover gases are used as fuel for the furnace, while the hydrogen is ready for the customers.

Industrial Piping and Equipment Estimating Manual. http://dx.doi.org/10.1016/B978-0-12-813946-2.00004-6

4.2   General Scope of Field Work Required for Each Steam Methane Reformer

96 MMSCFD SMR (Steam Methane Reformer)
Scope of Work-Field Erection
Radiant Section Erection

Set transfer line and outlet headers (including welding and hangers)
Set reformer lower supports and install floor panels
Install reformer wall panels
Install arch panels and temporary supports
Install reformer ladders, platforms, and stairs
Install burners
Install primary penthouse steel
Install reformer catalyst tubes
Install inlet piping system (including welding, spring hangers, etc.)
Install burner piping
Install finger ducts and down comers
Install penthouse secondary steel and siding
Install expansion joints and dampers
Miscellaneous (e.g., access doors, peep doors, catalyst tube boots)

Convection Section Erection

Set convection sections tower #1
Set transition duct (CC-4 to CC-5) and set CC-5 (APH), tower #1
Set ID fan inlet duct and CC-6 (APH), tower #2
Set transition ducts CC-6 to SCR (X-102) and set SCR (X-102) tower #2
Install crossover duct CC-6 to CC-5, install CC-5 (APH) outlet
FD Fan air outlet duct and FD fan cold air by-pass duct
Crossover duct tower #1 to tower #2 and transition duct radiant
Install expansion joints
Outlet duct ID fan and stack (including ladders and platforms)

Equipment Reformer Island

Set and assemble ID fan (C-103)
Set and assemble FD fan (C-102)
Install FD fan inlet duct and structure
Install X-135

Steam Drum Area

Set steam drum (V-108)
Set process gas boiler (E-103)
Install prereformer (V-103) including ladders and platforms
Miscellaneous Equipment
Set and assemble process air cooler (E-109) including steel
Set V-129 and V-107
Set deaerator (V-114) including ladder and platforms

Grouting

4.3   Steam Methane Reformer Estimate Data

4.3.1   Transfer Line and Outlet Headers, Lower Supports, and Floor Panels Sheet 1

Description	MH	Unit
Set Transfer Line and Outlet Headers (Including Welding and Hangers)		
Check foundation, AB, location, dimension, and elevation	24.00	JOB
Bush hammer foundation and set shims	1.63	EA
Set base plates and Teflon pads	1.50	EA
Set saddle w/base plates and support steel channel	1.00	EA
Set outlet transfer line sections; 38″ ID×30′10″/38″ ID×40′-9″	2.50	EA
Align sections and weld; 40″ OD×3/4″ AW 1-1/4 Cr-1/2 Mo	1.35	DI
Davit assembly, set outlet headers, manifold and spring hangers	8.00	EA
Erect/remove temporary wood support and temporary scaffold	16.00	EA
SH-15, SH-16, and SH-17	3.00	EA
Install structural calcium silicate block w/SS band	0.08	EA
Weld outlet pigtails (orbital) 1-1/4″ schedule xxs 800 HT	0.80	DI
Weld to outlet header 12″ ID×2″ MSW 20 CR-32 NI-Nb	3.88	DI
Weld to transfer line 22-1/2″ ID×1/2″ MSW 20 CR-32 NI-Nb	0.68	DI
Plywood covers/blow clean pigtails	0.10	EA
Set Reformer Lower Supports and Install Floor Panels		
Set base plates and pipe post	1.80	EA
Set structural and beams	14.00	ton
Connections	0.25	EA
Column splice joint 1″ diameter AB-32	1.50	EA
Make up AB 11/2″ diameter-56	0.29	EA
Set channel at erection joint (C15 back to back)	4.00	EA
Bolt channels together 3/4″ diameter bolts-408	0.08	EA
Set floor stiffener 34×4×11′-10″	1.00	EA
Connections	0.25	EA
Set floor plate	3.00	EA
Seal joints w/1/8″ stalastic	0.25	LF
Seal plate 12′×16″	4.00	EA

4.3.2 Wall/Arch Panels, Platforms, Burners, Penthouse Steel Sheet 2

Description	MH	Unit
Install Reformer Wall Panels		
Set wall panels	8.00	EA
Erection joint at walls and corners	16.00	EA
Bolt wall panels to floor structural	8.00	EA
Seal wall erection joints w/1/8″ stalastic	0.10	LF
Install Arch Panels and Temporary Supports		
Set arch support steel (back to back channels)	4.00	EA
Set arch panels	24.00	EA
Erection joints at floor and walls	16.00	EA
Square up and tighten floor, wall, and arch bolts	40.00	EA
Install/remove temporary support post/plug weld bolt holes	9.00	EA
Install tie rods w/turn buckles-level arch	3.00	EA
Install Reformer Ladders, Platforms, and Stairs		
Erect platforms	0.10	SF
Erect handrail w/toe plate attached	0.15	LF
Erection joint and connections	0.50	EA
Caged ladders	0.35	LF
KB	1.00	EA
Arch catwalk grating	0.15	SF
Install Burners		
Set gasket assembly on arch/angle frame	1.00	EA
Install burners and tile set	3.23	EA
Erect structural	6.50	ton
Install sag rods	0.79	EA
Connections-bolted and welded	0.50	EA

4.3.3 Catalyst Tubes, Inlet/Burner Pipe, Ducts/Downcomer, Penthouse Steel Sheet 4

Description	MH	Unit
Install Reformer Catalyst Tubes		
Temporary scaffold at radiant floor-erect/dismantle	40.00	JOB
Complete dimensional survey-panels and post	16.00	JOB
Install catalyst tubes through arch 3.5″ ID×45′-11-1/4″ 1287#/tube	1.50	EA
Install flexible expansion bellows on each tube	1.00	EA
Install catalyst support steel and catalyst hangers		
Install spring hanger support steel angle 5″×5″×1/4″	0.51	EA
Install SH-1	2.00	EA
Attach SH-1 to tubes	1.00	EA
Install Inlet Piping System (Including Welding, Spring Hangers, etc.)		
Install inlet subheaders into penthouse		
Install temporary wood blocking at pigtails	0.25	EA
Install inlet pipe 8″ schedule 100×24′ LG	8.00	EA
Install subheaders w/pigtails 8′ schedule 100/w 1″ sch 40	12.00	EA
Install spring hangers to inlet headers		
Install SH-8, SH-9, SH-10, SH-11, and SH-12	4.00	EA
Weld inlet pipe and subheaders		
8″ sch 100 (A312 Tp304H) BW	0.56	DI
Weld outlet pigtails (orbital) 1″ schedule 40 800 HT	1.40	DI
Install Burner Piping		
Erect/dismantle scaffold	40.00	JOB
Install pipe and header hangers SH-2, 3, 4, 5, 6, and 7	8.00	EA
24″ schedule 120 pipe	1.00	LF
8″ schedule 100 and 14″/16″ header	1.00	LF
Weld-24″ schedule 120 A312 TP304 BW	1.26	DI
Install Finger Ducts and Downcomers		
Install duct subheaders	10.00	EA
Install downcomer	1.00	LF
Bolted connection	0.15	EA
Bolt duct to structural	1.46	EA
Install Penthouse Secondary Steel and Siding		
Trim excess rod protrusions	40.00	JOB
Remove shipping braces	40.00	JOB
Girts	40.00	EA
Siding	0.25	SF
Cut opening for fuel line, manifold pipe and duct	1.00	JOB
Rigging scaffold at penthouse walls	48.00	JOB

4.3.4 Dampers, Modules, Transition Duct, Air Preheater Sheet 4

Description	MH	Unit
Install Expansion Joints and Dampers		
Dampers	32.00	EA
Expansion joints at duct	16.00	EA
Expansion joints at downcomers	2.00	LF
Miscellaneous (e.g., access doors, peep doors, catalyst tube boots)	6.00	EA
Set Convection Sections Tower #1		
Erect convection modules		
Set base plates	2.00	EA
Erect modules	64.00	EA
Erection joint 1-1/2″ batten glue—3/4″ bolt-1644	0.32	EA
Remove lifting lugs	0.50	EA
Remove tie rod and cpl'g	0.50	EA
Pack sleeve w/castable	1.00	EA
1″ pipe cap-screwed joint	0.45	EA
Bolt up at tube sheet bracket 7/8″ bolt-90	0.15	EA
Remove shipping steel and plywood	16.00	EA
Install access doors	4.00	EA
Bolted joint at header box	0.25	EA
Instrument TE, PT, FW	4.00	EA
12″ schedule 60 BW field weld	0.55	DI
16″ schedule 60 BW field weld	0.85	DI
12″ schedule 60 pipe	0.24	LF
20″ schedule 60 9Cr-1Mo-WP91 field	1.53	DI
22″ schedule 60 18-Cr-8Ni field weld	1.53	DI
Set Transition Duct (CC-4 to CC-5) and set CC-5 (APH), Tower #1		
Transition duct CC-4 to CC-5		
Set duct 7′×52′	32.00	EA
Erection joint at CC-5 1″ bolt	0.15	EA
Erect air preheater (CC-5)		
Unload and set SC17 and SC18	2.50	ton
Bolt up 3/4″ bolts	0.15	EA
Seal weld	0.35	LF
Set SC17/SC18 on support steel	1.25	ton
Bolt up to support steel 1-3/8″ bolt	0.15	EA
Unload and set SC1 through SC16	0.50	ton
Bolt blocks in pairs 3/4″ bolts	0.15	EA
Set blocks on support steel	2.50	ton
Bolt in place 3/4″ bolts	0.15	EA
Seal weld blocks		
Vertical/horizontal fillet weld between blocks	0.35	LF
Flat—blocks to support steel 3/16″ fillet weld	0.35	LF

4.3.5   Air Preheater, ID Fan Duct, Transition/Crossover Duct, Header Sheet 5

Description	MH	Unit
Erect air preheater (CC-6)		
Unload and set SC13 and SC14	2.50	ton
Bolt up 3/4″ bolts	0.15	EA
Seal weld	0.35	LF
Set SC13/SC14 on support steel	2.50	ton
Bolt up to support steel 1-3/8″ bolt	0.15	EA
Unload and set SC1 through SC12	0.50	ton
Bolt blocks in pairs 3/4″ bolts	0.15	EA
Set blocks on support steel	2.50	ton
Bolt in place 3/4″ bolts	0.15	EA
Seal weld blocks		
Vertical/horizontal fillet weld between blocks	0.35	LF
Flat—blocks to support steel 3/16″ fillet weld	0.35	LF
ID fan inlet duct		
Erect duct 10′×48′ and 10′×6′	10.00	EA
Erection joint 6 EA bolts	0.08	EA
Attach duct to support steel	2.33	EA
Set Transition Ducts CC-6 to SCR (X-102) and Set SCR (X-102) Tower #2		
Duct section 12′×52′ and 2′×48′	48.00	EA
Erection joint at SCR 1″ bolts	0.10	EA
Erection joint at CC-6 1″ bolts	0.15	EA
Bolt to structural steel	4.00	EA
SCR (x-102)		
Set unit reactor housing	64.00	EA
Untightened bolts/set required gaps/bolt up	48.00	EA
Monorail 3/4″ bolts	0.15	EA
Supporting beams 3/4″ bolts	0.15	EA
Bolt to structural 3/4″ bolts	0.15	EA
Install Crossover Duct CC-6 to CC-5, Install CC-5 (APH) Outlet		
Set duct CO-36833-11	24.00	EA
Erection joint 100″×197″	16.00	EA
Erection joint at SCR 1″ bolts	0.10	EA
Set duct to CC-5	16.00	EA
Erection joint 1″ bolts	0.10	EA
Combustion air duct header		
Duct sections 10′×10′×50′ and 8′×10′×50′	24.00	EA
Bolt duct to support steel	2.00	EA
Erection joint at inlet duct 1″ bolt	0.10	EA

4.3.6 FD Fan/By Pass Duct, Crossover/Transition Duct, Expansion Joints Sheet 6

Description	MH	Unit
FD Fan Air Outlet Duct and FD Fan Cold Air By-Pass Duct		
Erect duct 18′×10′ and 18′×8′	16.00	EA
Erection joint 3/4″ bolts	0.10	EA
Erection joint 4′-6″×4′-6″	8.00	EA
Erection joint 8′×8′	12.00	EA
Bolt duct to support steel	2.00	EA
FD fan by-pass duct	8.00	EA
Erect duct w/elbows 10′×4′-10″ and 20′×4′-10″ 2 elbows	20.00	EA
Erection joints 4′-10″×4′-10″	9.00	EA
Crossover Duct Tower #1 to Tower #2 and Transition Duct Radiant		
Crossover duct tower #1 to #2		
Erect duct section 9′×9′×60′	24.00	EA
Erection joint 3/4″ bolts	0.10	EA
Bolt duct to steel	2.00	EA
Erection joint-assemble duct sections	12.00	EA
Crossover duct SCR to CC-5		
Erect duct 16′-5″×42′	24.00	EA
Erection joints at SCR and CC-5	16.00	EA
Set duct CO-36833-10	16.00	EA
Erection joint at elbow	16.00	EA
Bolt to support steel	2.00	EA
Transition radiant to convection–combustion air duct		
Remove angle 6×31/2 flange 8′-3″×2′-3″ 2 EA and 8′-3″×3′-0″ 7 EA	0.50	EA
Weld flange to duct	0.35	EA
Assemble duct 6 sections	8.00	EA
Field splice	0.35	EA
Set duct header	32.00	EA
Bolt duct to support steel	2.00	EA
Erection joint at convection duct 3/4″ bolts	0.10	EA
Install expansion joints	4.00	EA
Make up expansion joints	10.00	EA
Expansion joint 4 113″×485″	8.00	EA

4.3.7 Outlet Duct, Flue Gas Stack, ID and FD Fans Sheet 7

Description	MH	Unit
Outlet duct ID fan and stack (including ladders and platforms)		
Erect duct 10′×48′ and 10′×40′ and elbow	24.00	EA
Erection joint at elbows	16.00	EA
Flue gas outlet duct to CC-6 1″ bolts	0.10	EA
Spool connection to CC-6 1″ bolts	0.10	EA
Damper D-1	51.00	EA
Flue gas stack		
Erect lower section 10′ ID×59′-6″	2.50	ton
Attach lower section to base foundation 21/4″ bolts	0.45	EA
Erect upper section 10′ ID×59′-6″	2.50	ton
Erection joint 3/4″ bolts w/1-1/2″ thk×1-1/2″ wide batten glued	0.35	EA
Ladders and platforms	32.00	EA
Remove lifting lugs	2.00	EA
Erect/dismantle scaffold	16.00	JOB
Remove shipping steel	16.00	JOB
Equipment Reformer Island		
Set and Assemble ID Fan (C-103) and FD Fan (C-104)		
ID fan		
Install sole and base plates	2.00	EA
Set fan housing-split	12.00	EA
Set impeller assembly/shaft	16.00	EA
Alignment fan bearings	40.00	EA
Seal weld housing inside/outside	48.00	JOB
Weld inlet cones and stud rings	12.00	JOB
Weld shaft seal stud rings	12.00	JOB
Coupling alignment	32.00	JOB
Install damper	40.00	EA
Attach fan to foundation 15/8″ bolts-12 1″ bolts	0.42	EA

4.3.8   FD Fan, NH3 Skid, Steam Drum, Gas Boiler, Prereformer Sheet 8

Description	MH	Unit
Install FD Fan Inlet Duct and Structure		
Erect duct	32.00	EA
Erection joint	16.00	EA
Set support steel	8.00	EA
Structure	40.00	EA
Install X-135		
Set X-135-NH3 vaporizer skid 9'-10″ × 16' × 19'-8″	58.00	EA
Attach to foundation 3/4″ bolts	0.25	EA
Steam Drum Area		
Set Steam Drum (V-108)		
Set Teflon pad and shim	4.00	EA
Bolt pad to structural	4.00	EA
Set drum 159000#	2.50	ton
Open/close man ways for inspection	12.00	EA
Set Process Gas Boiler (E-103)		
Set slide plates and fixed/sliding saddle	4.00	EA
Attach saddles to foundation	4.00	EA
Set E-103 84″ ID × 44' 269,000#	2.50	ton
Stud bolt assemblies and fit up ring/remove	4.00	EA
Field weld to transfer line 40″ OD × 3/4″AW 1-1/4Cr-1/2Mo	51.52	EA
Install Prereformer, Cooler, and Deaerator Including Ladders and Platforms		
Set skirt	16.00	EA
Set prereformer, coolers, and deaerator	2.50	ton
Attach to foundation	8.00	EA
Platform	0.10	SF
Grating	0.15	SF
Handrail	0.25	LF
Connection	0.50	EA
Caged ladder	0.35	LF

4.3.9   Process Air Cooler, BD Drums, Deaerator Sheet 9

Description	MH	Unit
Set & Assemble Process Air Cooler (E-109) Including Steel		
Set coolers 11'-6" × 28' 31,700#	2.50	ton
Assemble braces	2.00	EA
Connections	0.50	EA
Walkway & header support	4.00	EA
Platform	0.10	SF
Grating	0.15	SF
Handrail	0.15	LF
Continuous & intermittent BD drums 2'-6" OD × 9' T/T 2000#	16.00	EA
Set Deaerator (V-114) Including Ladder and Platforms		
Set V-114 10' OD × 20'-4" TL/TL 26,500#	2.50	ton
Bolt to foundation	2.00	EA
Open/close man ways for inspection	4.00	EA
Platform	0.10	SF
Grating	0.15	SF
Handrail	0.15	LF
Connections	0.50	LF
Grouting	780.00	UNIT

4.4 Steam Methane Reformer Installation Estimate

4.4.1 Transfer Line and Outlet Headers, Lower Supports, and Floor Panels Sheet 1

Description	MH	Historical Quantity	Unit	Estimate Quantity	Unit	BM
Set transfer line and outlet headers (include welding and hangers)						**0**
Check foundation, AB, location, dimension, and elevation	24.00	1.0	JOB	0	JOB	0
Bush hammer foundation and set shims	1.63	9.0	EA	0	EA	0
Set base plates and Teflon pads	1.5	8.0	EA	0	EA	0
Set saddle w/base plates	1.0	20.0	EA	0	EA	0
Set outlet transfer line sections; 38″ ID×30′-10″/38″ ID× 40′-9″	2.5	34.6	EA	0	EA	0
Align sections and weld; 40″ OD×3/4″ AW 1-1/4 Cr-1/2 Mo	1.35	40.0	DI	0	DI	0
Install davit assembly	8.00	1.0	EA	0	EA	0
Set outlet headers w/pigtails attached 8-1/2″ ID×19-8″ LG	8.0	8.0	EA	0	EA	0
Erect/remove temporary wood support	16.0	1.0	EA	0	EA	0
At outlet manifold R-1 spring hanger 8-1/2″ ID pipe	8.0	8.0	EA	0	EA	0
SH-14 2 EA-spring hanger w/C6×8.2	8.0	8.0	EA	0	EA	0
SH-15, SH-16 and SH-17	3.0	128.0	EA	0	EA	0
Support steel C8×11.5 and C6×8.2	1.0	64.0	EA	0	EA	0
Install structural calcium silicate block w/SS band	0.08	768.0	EA	0	EA	0
Temporary support scaffold at outlet-erect/remove	16.0	1.0	EA	0	EA	0
Weld outlet pigtails (orbital) 1-1/4″ schedule xxs 800 HT	0.8	960.0	DI	0	DI	0
Weld to outlet header 12″ ID× 2″ MSW 20CR-32NI-Nb	3.88	48.0	DI	0	DI	0
Weld to transfer line 22-1/2″ ID× 1/2″ MSW 20CR-32NI-Nb	0.68	90.0	DI	0	DI	0
Plywood covers/blow clean pigtails	0.1	384.0	EA	0	EA	0
Set Reformer Lower Supports and Install						**0**
Floor Panels						
Set structural steel post						
Set base plates	1.8	35.0	EA	0	EA	0
Set structural	14.0	15.0	ton	0	ton	0
Connections	0.25	54.0	EA	0	EA	0
Set 6″ diameter pipe post 7′ LG	4.0	20.0	EA	0	EA	0
Column splice joint 1″ diameter AB	1.5	20.0	EA	0	EA	0
Make up AB 1-1/2″ diameter	0.29	160.0	EA	0	EA	0
Set perimeter beam	14.0	5.0	ton	0	ton	0
Connections	0.25	20.0	EA	0	EA	0
Set channel at erection joint (C15 back to back)	4.0	3.0	EA	0	EA	0
Bolt channels together 3/4″ diameter bolts	0.08	408.0	EA	0	EA	0
Set floor stiffener 34×4×11′-10″	1.0	136.0	EA	0	EA	0
Connections	0.25	272.0	EA	0	EA	0
Set floor plate	3.0	36.0	EA	0	EA	0
Seal joints w/1/8″ stalastic	0.25	189.0	LF	0	LF	0
Seal plate 12′×16″	4.0	32.0	EA	0	EA	0

4.4.2 Wall, Arch Panels, Platforms, Burners, Penthouse Steel Sheet 2

Description	MH	Historical Quantity	Unit	Estimate Quantity	Unit	BM
Install Reformer Wall Panels						0
Set wall panels	8.00	20.0	EA	0	EA	0
Erection joint at walls and corners	16.00	20.0	EA	0	EA	0
Bolt wall panels to floor structural	8.00	20.0	EA	0	EA	0
Seal wall erection joints w/1/8″ stalastic	0.10	720.0	LF	0	LF	0
Install Arch Panels and Temporary Supports						0
Set arch support steel (back to back channels)	4.00	6.0	EA	0	EA	0
Set arch panels	24.00	5.0	EA	0	EA	0
Erection joints at floor and walls	16.00	6.0	EA	0	EA	0
Seal joints w/1/8″ stalastic	0.25	270.0	LF	0	LF	0
Square up and tighten floor, wall, and arch bolts	40.00	1.0	EA	0	EA	0
Install/remove temporary support post	9.00	24.0	EA	0	EA	0
Install tie rods w/turn buckles-level arch	3.00	24.0	EA	0	EA	0
Install Reformer Ladders, Platforms and Stairs						0
Erect platforms #1, #2, and #3	0.10	2160.0	SF	0	SF	0
Erect handrail w/toe plate attached	0.15	1440.0	LF	0	LF	0
Erection joint and connections	0.50	60.0	EA	0	EA	0
Caged ladders	0.35	103.0	LF	0	LF	0
KB	1.00	60.0	EA	0	EA	0
Arch catwalk grating	0.15	1875.0	SF	0	SF	0
Erect handrail w/toe plate attached	0.15	146.0	LF	0	LF	0
Caged ladders	0.35	20.0	LF	0	LF	0
Erection joint and connections	0.50	10.0	EA	0	EA	0
Platform #4	0.10	272.0	SF	0	SF	0
Erect handrail w/toe plate attached	0.15	150.0	LF	0	LF	0
KB	1.00	6.0	EA	0	EA	0
Caged ladders	0.35	56.0	LF	0	LF	0
Erection joint and connections	0.50	10.0	EA	0	EA	0
Platform #10, 11, 12, 13, 14, 14A	0.10	812.0	SF	0	SF	0
Erect handrail w/toe plate attached	0.15	468.0	LF	0	LF	0
Caged ladders	0.35	120.0	LF	0	LF	0
Erection joint and connections	0.50	20.0	EA	0	EA	0
Install Burners						0
Set gasket assembly on arch/angle frame	1.00	104.0	EA	0	EA	0
Install burners and tile set	3.23	104.0	EA	0	EA	0
Install Primary Penthouse Steel						0
Erect structural	6.50	110.0	ton	0	ton	0
Connections bolted and welded	0.50	200.0	EA	0	EA	0
Install sag rods	0.79	32.0	EA	0	EA	0

4.4.3   Catalyst Tubes, Inlet/Burner Pipe, Ducts/Downcomer, Penthouse Steel Sheet 4

		Historical			Estimate		
Description	MH	Quantity	Unit	Quantity	Unit	BM
Install Reformer Catalyst Tubes						0
Temporary scaffold at radiant floor erect/dismantle	40.00	1	JOB	0	JOB	0
Complete dimensional survey panels and post	16.00	1	JOB	0	JOB	0
Install catalyst tubes through arch 3.5″ ID×45′-11-1/4″ 1287#/tube	1.50	384	EA	0	EA	0
Install flexible expansion bellows on each tube	1.00	384	EA	0	EA	0
Install catalyst support steel and catalyst hangers						
Install spring hanger support steel angle 5″×5″×1/4″	0.51	211	EA	0	EA	0
Install SH-1	2.00	192	EA	0	EA	0
Attach SH-1 to tubes	1.00	192	EA	0	EA	0
Install Inlet Piping System						0
(Including Welding, SH)						
Install inlet subheaders into penthouse						
Install temporary wood blocking at pigtails	0.25	384	EA	0	EA	0
Install inlet pipe 8″ schedule 100×24′ LG	8.00	8	EA	0	EA	0
Install subheaders w/pigtails 8′ schedule 100/w 1″ schedule 40	12.00	16	EA	0	EA	0
Install spring hangers to inlet headers						
Install SH-8, SH-9, SH-10, SH-11, and SH-12	4.00	96	EA	0	EA	0
Weld inlet pipe and subheaders						
8″ Schedule 100 (A312 Tp304H) BW	0.56	320	DI	0	DI	0
Weld outlet pigtails (orbital) 1″ schedule 40 800 HT	1.40	384	DI	0	DI	0
Install Burner Piping						0
Install fuel gas pipe and headers						
Erect/dismantle scaffold	40.00	1	JOB	0	JOB	0
Install pipe and header hangers SH-2, 3, 4, 5, 6, and 7	8.00	9	EA	0	EA	0
Hang pipe and headers from hangers						
24″ schedule 120 pipe	1.00	180	LF	0	LF	0
8″ schedule 100 and 14″/16″ header	1.00	56	LF	0	LF	0
Weld-24″ schedule 120 A312 TP304 BW	1.26	120	DI	0	DI	0
Install Finger Ducts and Down comers						0
Install duct subheaders	10.00	9	EA	0	EA	0
Down comer 18″ diameter	1.00	91	LF	0	LF	0
Down comer 14″ diameter	1.00	26	LF	0	LF	0
Bolt down comer to duct subheaders	0.15	936	EA	0	EA	0
Bolt duct to structural	1.46	36	EA	0	EA	0
Install Penthouse Secondary Steel and Siding						0
Trim excess rod protrusions	40.00	1	JOB	0	JOB	0
Remove shipping braces	40.00	1	JOB	0	JOB	0
Grits	40.00	3	EA	0	EA	0
Siding	0.25	3622	SF	0	SF	0
Cut opening for fuel line, manifold pipe, and duct	1.00	14	JOB	0	JOB	0
Rigging scaffold at penthouse walls	48.00	1	JOB	0	JOB	0

4.4.4 Dampers, Modules, Transition Duct, Air Preheater Sheet 4

		Historical		Estimate		
Description	MH	Quantity	Unit	Quantity	Unit	BM
Install Expansion Joints and Dampers						0
Dampers	32.00	9.0	EA	0	EA	0
Expansion joints at duct	16.00	9.0	EA	0	EA	0
Expansion joints at 18" down comer	2.00	91.0	LF	0	LF	0
Expansion joints at 14" down comer	2.00	26.0	LF	0	LF	0
Miscellaneous (e.g., access doors, peep doors)	6.00	16.0	EA	0	EA	0
Set Convection Sections Tower #1						0
Erect convection modules						
Set base plates	2.00	13.0	EA	0	EA	0
Erect modules	64.00	6.0	EA	0	EA	0
Erection joint 1-1/2" batten glue—3/4" bolt	0.32	756.0	EA	0	EA	0
Remove lifting lugs	0.50	20.0	EA	0	EA	0
Remove tie rod and coupling	0.50	23.0	EA	0	EA	0
Pack sleeve w/castable	1.00	23.0	EA	0	EA	0
1" pipe cap-screwed joint	0.45	23.0	EA	0	EA	0
Bolt up at tube sheet bracket 7/8" bolt	0.15	90.0	EA	0	EA	0
Remove shipping steel and plywood	16.00	1.0	EA	0	EA	0
Install access doors	4.00	2.0	EA	0	EA	0
Bolted joint at header box	0.25	20.0	EA	0	EA	0
Instrument TE, PT, FW	4.00	3.0	EA	0	EA	0
Field welds at CC-4						
12" schedule 60 BW	0.55	12.0	DI	0	DI	0
16" schedule 60 BW	0.85	32.0	DI	0	DI	0
12" schedule 60 pipe	0.24	20.0	LF	0	LF	0
Field welds at CC-3						
20" schedule 60 9Cr-1Mo-WP91	1.53	20.0	DI	0	DI	0
22" schedule 60 18-Cr-8Ni	1.53	44.0	DI	0	DI	0
Set Transition Duct (CC-4) and Set CC-5, Tower #1						0
Transition duct CC-4 to CC-5						
Set duct 7'×52'	32.00	1.0	EA	0	EA	0
Erection joint at CC-5 1" bolt	0.15	226.0	EA	0	EA	0
Erect air preheater (CC-5)						
Unload and set SC17 and SC18	2.50	35.0	ton	0	ton	0
Bolt up 3/4" bolts	0.15	64.0	EA	0	EA	0
Seal weld	0.35	104.0	LF	0	LF	0
Set SC17/SC18 on support steel	1.25	35.0	ton	0	ton	0
Bolt up to support steel 1-3/8" bolt	0.15	64.0	EA	0	EA	0
Unload and set SC1 through SC16	0.50	208.0	ton	0	ton	0
Bolt blocks in pairs 3/4" bolts	0.15	448.0	EA	0	EA	0
Set blocks on support steel	2.50	26.5	ton	0	ton	0
Bolt in place 3/4" bolts	0.15	128.0	EA	0	EA	0
Seal weld blocks						
Vertical/horizontal fillet weld between blocks	0.35	398.0	LF	0	LF	0
Flat blocks to support steel 3/16" fillet weld	0.35	112.0	LF	0	LF	0

4.4.5   Air Preheater, ID Fan Duct, Transition/Crossover Duct, Header Sheet 5

Description	MH	Historical Quantity	Unit	Estimate Quantity	Unit	BM
Set ID Fan Inlet Duct and CC-6 (APH),						**0**
Tower #2						
Erect air preheater (CC-6)						
Unload and set SC13 and SC14	2.50	25.8	ton	0	ton	0
Bolt up 3/4″ bolts	0.15	42.0	EA	0	EA	0
Seal weld	0.35	126.0	LF	0	LF	0
Set SC13/SC14 on support steel	2.50	25.8	ton	0	ton	0
Bolt up to support steel 1-3/8″ bolt	0.15	64.0	EA	0	EA	0
Unload and set SC1 through SC12	0.50	252.0	ton	0	ton	0
Bolt blocks in pairs 3/4″ bolts	0.15	336.0	EA	0	EA	0
Set blocks on support steel	2.50	42.0	ton	0	ton	0
Bolt in place 3/4″ bolts	0.15	96.0	EA	0	EA	0
Seal weld blocks						
Vertical/horizontal fillet weld between	0.35	300.0	LF	0	LF	0
blocks						
Flat blocks to support steel 3/16″ fillet weld	0.35	98.0	LF	0	LF	0
ID fan inlet duct						
Erect duct 10′×48′ and 10′×6′	10.00	2.0	EA	0	EA	0
Erection joint 6 EA bolts	0.08	1000.0	EA	0	EA	0
Attach duct to support steel	2.33	6.0	EA	0	EA	0
Set Transition Ducts CC-6 and Set SCR						**0**
(X-102) Tower #2						
Transition duct X-102 to CC-6						
Duct section 12′×52′ and 2′×48′	48.00	2.0	EA	0	EA	0
Erection joint at SCR 1″ bolts	0.10	164.0	EA	0	EA	0
Erection joint at CC-6 1″ bolts	0.15	204.0	EA	0	EA	0
Bolt to structural steel	4.00	8.0	EA	0	EA	0
SCR (x-102)						
Set unit reactor housing	64.00	1.0	EA	0	EA	0
Untightened bolts/set required gaps/bolt up	48.00	1.0	EA	0	EA	0
Monorail 3/4″ bolts	0.15	20.0	EA	0	EA	0
Supporting beams 3/4″ bolts	0.15	20.0	EA	0	EA	0
Bolt to structural 3/4″ bolts	0.15	40.0	EA	0	EA	0
Install Crossover Duct CC-6, Install CC-5						**0**
Outlet						
Combustion air duct header						
Transition crossover duct						
Set duct CO-36833-11	24.00	1.0	EA	0	EA	0
Erection joint 100″×197″	16.00	1.0	EA	0	EA	0
Erection joint at SCR 1″ bolts	0.10	188.0	EA	0	EA	0
Set duct to CC-5	16.00	1.0	EA	0	EA	0
Erection joint 1″ bolts	0.10	240.0	EA	0	EA	0
Combustion air duct header						
Duct sections 10′×10′×50′ and 8′×10′×50′	24.00	2.0	EA	0	EA	0
Bolt duct to support steel	2.00	13.0	EA	0	EA	0
Erection joint at inlet duct 1″ bolt	0.10	226.0	EA	0	EA	0

4.4.6 FD Fan/Duct, Crossover/Transition Duct, Expansion Joints Sheet 5

Description		Historical			Estimate		
	MH	Quantity	Unit	Quantity	Unit	BM
FD Fan Air Outlet Duct and FD Fan Cold							0
Air By-Pass Duct							
FD fan outlet duct							
Erect duct 18′×10′ and 18′×8′	16.00	2.0	EA	0	EA	0
Erection joint 3/4″ bolts	0.10	64.0	EA	0	EA	0
Erection joint 4′-6″×4′-6″	8.00	1.0	EA	0	EA	0
Erection joint 8′×8′	12.00	1.0	EA	0	EA	0
Bolt duct to support steel	2.00	2.0	EA	0	EA	0
Set support steel							
FD fan by-pass duct	8.00	1.0	EA	0	EA	0
Erect duct w/elbows 10′×4-10″ and 20′×4′-10″	20.00	2.0	EA	0	EA	0
Erection joints 4′-10″x4′-10″	9.00	1.0	EA	0	EA	0
Crossover Duct Tower #1 to Tower #2							0
Section to convection section							
Crossover duct tower #1 to #2							
Erect duct section 9′×9′×60′	24.00	1.0	EA	0	EA	0
Erection joint 3/4″ bolts	0.10	228.0	EA	0	EA	0
Bolt duct to steel	2.00	8.0	EA	0	EA	0
Erection joint-assemble duct sections	12.00	1.0	EA	0	EA	0
Crossover duct SCR to CC-5							
Erect duct 16′-5″×42′	24.00	1.0	EA	0	EA	0
Erection joints at SCR and CC-5	16.00	2.0	EA	0	EA	0
Set duct CO-36833-10	16.00	1.0	EA	0	EA	0
Erection joint at elbow	16.00	1.0	EA	0	EA	0
Bolt to support steel	2.00	1.0	EA	0	EA	0
Transition radiant to combustion air duct							
Remove angle 6×31/2 flange	0.50	9.0	EA	0	EA	0
Weld flange to duct	0.35	207.0	EA	0	EA	0
Assemble duct 6 sections	8.00	6.0	EA	0	EA	0
Field splice	0.35	172.0	EA	0	EA	0
Set duct header	32.00	1.0	EA	0	EA	0
Bolt duct to support steel	2.00	8.0	EA	0	EA	0
Erection joint at convection duct 3/4″ bolts	0.10	96.0	EA	0	EA	0
Install Expansion Joints							0
Expansion joint 11 76″×218″	4.00	2.0	EA	0	EA	0
Expansion joint 8 120″×120″	4.00	1.0	EA	0	EA	0
Expansion joint 14 78″×268.5″	4.00	2.0	EA	0	EA	0
Expansion joint 13 76″×266.5″	4.00	2.0	EA	0	EA	0
Expansion joint 12 111″×111″	4.00	1.0	EA	0	EA	0
Expansion joints 2 and 3 101″×583″ and	4.00	2.0	EA	0	EA	0
100″×197″							
Expansion joint 1 101″×583″	4.00	1.0	EA	0	EA	0
Expansion joints 7 and 9 71.5″×64″ and	4.00	2.0	EA	0	EA	0
67.25″×95″							
Expansion joint 10 35.7″×72″	4.00	2.0	EA	0	EA	0
Expansion joint 19 58″×54″	4.00	1.0	EA	0	EA	0
Expansion joints 5 and 6 118″×490″ and	4.00	2.0	EA	0	EA	0
43″×124.5″							
Make up expansion joints	10.00	18.0	EA	0	EA	0
Expansion joint 4 113′×485″	8.00	1.0	EA	0	EA	0

4.4.7 Outlet Duct, Flue Gas Stack, ID and FD Fans Sheet 7

Description	Historical			Estimate		
	MH	Quantity	Unit	Quantity	Unit	BM
Outlet Duct ID Fan and Stack						0
(Including Platforms)						
Outlet duct						
Erect duct 10′×48′ and 10′×40′ and elbow	24.00	3.0	EA	0	EA	0
Erection joint at elbows	16.00	3.0	EA	0	EA	0
Flue gas outlet duct to CC-6 1″ bolts	0.10	194.0	EA	0	EA	0
Spool connection to CC-6 1″ bolts	0.10	96.0	EA	0	EA	0
Damper D-1	51.00	1.0	EA	0	EA	0
Flue gas stack						
Erect lower section 10′ ID×59′-6″	2.50	32.8	ton	0	ton	0
Attach lower section to base foundation	0.45	72.0	EA	0	EA	0
Erect upper section 10′ ID×59′-6″	2.50	25.5	ton	0	ton	0
Erection joint 3/4″ bolts	0.35	44.0	EA	0	EA	0
Ladders and platforms	32.00	1.0	EA	0	EA	0
Remove lifting lugs	2.00	2.0	EA	0	EA	0
Erect/dismantle scaffold	16.00	1.0	JOB	0	JOB	0
Remove shipping steel	16.00	1.0	JOB	0	JOB	0
Equipment Reformer Island						
Set and Assemble ID Fan						0
(C-103)						
ID fan						
Install sole and base plates	2.00	6.0	EA	0	EA	0
Set fan housing-split	12.00	2.0	EA	0	EA	0
Set impeller assembly/shaft	16.00	1.0	EA	0	EA	0
Alignment fan bearings	40.00	1.0	EA	0	EA	0
Seal weld housing inside/outside	48.00	1.0	JOB	0	JOB	0
Weld inlet cones and stud rings	12.00	1.0	JOB	0	JOB	0
weld shaft seal stud rings	12.00	1.0	JOB	0	JOB	0
coupling alignment	32.00	1.0	JOB	0	JOB	0
Install damper	40.00	1.0	EA	0	EA	0
attach fan to foundation	0.42	24.0	EA	0	EA	0
Set and Assemble FD Fan						0
(C-102)						
FD fan						
Install sole and base plates	2.00	6.0	EA	0	EA	0
Set fan housing split	12.00	2.0	EA	0	EA	0
Set impeller assembly/shaft	16.00	1.0	EA	0	EA	0
Alignment fan bearings	40.00	1.0	EA	0	EA	0
Seal weld housing inside/outside	48.00	1.0	JOB	0	JOB	0
Weld inlet cones and stud rings	12.00	1.0	JOB	0	JOB	0
Weld shaft seal stud rings	12.00	1.0	JOB	0	JOB	0
Coupling alignment	32.00	1.0	EA	0	EA	0
Install damper	40.00	1.0	EA	0	EA	0
Attach fan to foundation	0.42	24.0	EA	0	EA	0

4.4.8 FD Fan, NH3 Skid, Steam Drum, Gas Boiler, Prereformer Sheet 8

Description	Historical			Estimate		
	MH	Quantity	Unit	Quantity	Unit	BM
Install FD Fan Inlet Duct and Structure				0		**0**
Erect duct	32.00	2.0	EA	0	EA	0
Erection joint	16.00	1.0	EA	0	EA	0
Set support steel	8.00	1.0	EA	0	EA	0
Structure	40.00	1.0	EA	0	EA	0
Install X-135						**0**
Set X-135-NH3 vaporizer skid 9'-10"×16'×19'-8"	58.00	1.0	EA	0	EA	0
Attach to foundation 3/4" bolts	0.25	6.0	EA	0	EA	0
Steam Drum Area						
Set Steam Drum (V-108)						**0**
Set Teflon pad and shim	4.00	2.0	EA	0	EA	0
Bolt pad to structural	4.00	2.0	EA	0	EA	0
Set drum 159,000#	2.50	79.5	ton	0	ton	0
Open/close man ways for inspection	12.00	2.0	EA	0	EA	0
Set Process Gas Boiler (E-103)						**0**
Set slide plates	4.00	4.0	EA	0	EA	0
Set fixed/sliding saddle	4.00	2.0	EA	0	EA	0
Attach saddles to foundation	4.00	2.0	EA	0	EA	0
Set E-103 84" ID×44' 269,000#	2.50	134.5	ton	0	ton	0
Stud bolt assemblies and fit up ring/remove	4.00	2.0	EA	0	EA	0
Field weld to transfer line 40" OD×3/4"AW 1-1/4Cr-1/2Mo	51.52	1.0	EA	0	EA	0
Install Prereformer (V-103) Including Ladders and Platforms						**0**
Set skirt	16.00	1.0	EA	0	EA	0
Set V-103 103" ID×22'-4" T/T	2.50	48.8	ton	0	ton	0
Attach to foundation	8.00	1.0	EA	0	EA	0
Platform	0.10	170.0	SF	0	SF	0
Grating	0.15	170.0	SF	0	SF	0
Handrail	0.25	72.0	LF	0	LF	0
Connection	0.50	16.0	EA	0	EA	0
Caged ladder	0.35	40.0	LF	0	LF	0

4.4.9   Process Air Cooler, BD Drums, Deaerator Sheet 9

Description	Historical			Estimate		
	MH	Quantity	Unit	Quantity	Unit	BM
Miscellaneous Equipment						
Set and Assemble Process Air Cooler (E-109) Including Steel						**0**
Set coolers 11'-6"×28' 31,700#	2.50	15.9	ton	0	ton	0
Assemble braces	2.00	24.0	EA	0	EA	0
Connections	0.50	48.0	EA	0	EA	0
Walkway and header support	4.00	2.0	EA	0	EA	0
Platform	0.10	196.0	SF	0	SF	0
Grating	0.15	196.0	SF	0	SF	0
Handrail	0.15	98.0	LF	0	LF	0
Set V-129 and V-107						**0**
Continuous and intermittent BD drums 2'-6" OD×9' T/T	16.00	2.0	EA	0	EA	0
Set Deaerator (V-114) Including Ladder and Platforms						**0**
Set V-114 10' OD×20'-4" TL/TL 26,500#	2.50	13.3	ton	0	ton	0
Bolt to foundation	2.00	4.0	EA	0	EA	0
Open/close man ways for inspection	4.00	2.0	EA	0	EA	0
Platform	0.10	70.0	SF	0	SF	0
Grating	0.15	70.0	SF	0	SF	0
Handrail	0.15	32.0	LF	0	LF	0
Connections	0.50	12.0	LF	0	LF	0
Grouting	780.00	1.0	UNIT	0	UNIT	0

4.5   96 MMSCFD Steam Methane Reformer—Equipment Installation Man Hours

	Actual	Estimated
Facility—Refinery and Hydrogen Plant Equipment	**MH**	**BM**
Set transfer line and outlet headers (includes welding and hangers)	2006	0
Set reformer lower supports and install floor panels	1049	0
Install reformer wall panels	712	0
Install arch panels and temporary supports	636	0
Install reformer ladders, platforms, and stairs	1157	0
Install burners	440	0
Install primary penthouse steel	840	0
Install reformer catalyst tubes	1700	0
Install inlet piping system (including welding, spring hangers, etc.)	1453	0
Install burner piping	499	0
Install finger ducts and downcomers	400	0
Install penthouse secondary steel and siding	1168	0
Install expansion joints and dampers	666	0
Miscellaneous (e.g., access doors, peep doors, catalyst tube boots)	96	0
Set convection sections tower #1	900	0
Set transition duct (CC-4 to CC-5) and set CC-5 (APH), tower #1	688	0
Set ID fan inlet duct and CC-6 (APH), tower #2	738	0
Set transition ducts CC-6 to SCR (X-102) and set SCR (X-102) tower #2	300	0
Install crossover duct CC-6 to CC-5, install CC-5 (APH) outlet	195	0
FD Fan air outlet duct and FD fan cold air by-pass duct	119	0
Crossover duct tower #1 to tower #2 and transition duct radiant	408	0
Install expansion joints	260	0
Outlet duct ID fan and stack (including ladders and platforms)	462	0
Set and assemble ID fan (C-103)	246	0
Set and assemble FD fan (C-102)	246	0
Install FD fan inlet duct and structure	128	0
Install X-135	60	0
Set steam drum (V-108)	239	0
Set process gas boiler (E-103)	428	0
Install prereformer (V-103) including ladders and platforms	228	0
Set and assemble process air cooler (E-109) including steel	183	0
Set V-129 and V-107	32	0
Set deaerator (V-114) including ladder and platforms	77	0
Grouting	780	0
Equipment installation man hours	**19,538**	**0**

4.6   General Scope of Field Work Required for Each Pressure Swing Adsorbers (PSA)

Pressure Swing Adsorbers (PSA)
  Scope of Work-Field Erection

Unloading, receipt, and storage of material and equipment
Foundation check
Vessel and skid erection
Offload adsorbers, skids, and surge vessels
Assemble-install screen in adsorbers six sections/vessel
Shim base ring adsorbers and surge drums
Set adsorbers
Set surge vessels
Set PSA skids
Set adsorbers
Interconnecting piping; skid and vessel IC piping
Adsorber filling

4.7   Pressure Swing Adsorber Estimate Data

4.7.1   Receive, Erect Equipment, IC Skids, and Vessel Piping Sheet 1

Description	MH	Unit
Unloading, Receipt, and Storage of Material and Equipment		
Equipment Erection		
Offload adsorbers, skids, and surge vessels	10.00	EA
Assemble-install screen in adsorbers six sections/vessel	2.00	EA
Shim base ring adsorbers and surge drums	3.00	EA
Anchor bolts 1-5/8″	0.25	EA
PSA Area		
Set adsorbers and surge vessels	1.80	ton
Set PSA skids	2.50	ton
Install Interconnecting Skid and Vessel Piping		
1″ std pipe, sw, valve, bolt up	1.97	LF
1″ std pipe, sw, valve, bolt up	0.16	LF
3″ std pipe, bw, valve, bolt up	0.31	LF
4″ std pipe, bw, valve, bolt up	0.40	LF
6″ std pipe, bw, valve, bolt up	0.67	LF
8″ std pipe, bw, valve, bolt up	0.89	LF
14″ std pipe, bw, valve, bolt up	1.52	LF
18″ std pipe, bw, valve, bolt up	2.03	LF
20″ std pipe, bw, valve, bolt up	6.53	LF
24′ std pipe, bw, valve, bolt up	2.14	LF
30″ std pipe, bw, valve, bolt up	3.13	LF
24′ std pipe, bw, valve, bolt up	2.40	LF
Hydro test pipe	0.12	LF

4.8 Pressure Swing Adsorber Equipment Installation Estimate

4.8.1 Receive, Erect Equipment, IC Skids, and Vessel Piping Sheet 1

Description	Historical			Estimate			
	MH	Quantity	Unit	Quantity	Unit	BM	PF
Unloading, Receipt, and Storage of Material and Equipment	380.00	1.0	Unit	0	Unit	**0**	
Equipment Erection						**0**	
Offload adsorbers, skids, and surge vessels	10.00	13.0	EA	0	EA	**0**	
Assemble-install screen in adsorbers six sections/vessel	2.00	60.0	EA	0	EA	**0**	
Shim base ring adsorbers and surge drums	3.00	12.0	EA	0	EA	**0**	
Anchor bolts 1-5/8"—16 EA at adsorbers	0.25	160.0	EA	0	EA	**0**	
Anchor bolts 1-5/8"—24 EA at surge drums	0.25	54.0	EA	0	EA	**0**	
PSA Area						**0**	
Set adsorbers V-8000 A, C, E, H, and J (12′ ID×45′) 192,400#	1.80	96.2	ton	0	ton	**0**	
Set surge vessels V-113 A and B (12′ ID×100′) 154000#	1.80	77.0	ton	0	ton	**0**	
Set PSA skids X-101 A, B, and C (50′×15′×47′) 75,000#	2.50	37.9	ton	0	ton	**0**	
Set adsorbers V-8000 B, D, F, G, and I (12′ ID×45′) 192,400#	1.80	96.2	ton	0	ton	**0**	
Install Interconnecting Skid and Vessel Piping							**0**
1″ std pipe, sw, valve, bolt up	1.97	81.0	LF	0	LF		0
1″ std pipe, sw, valve, bolt up	0.16	207.0	LF	0	LF		0
3″ std pipe, bw, valve, bolt up	0.31	95.0	LF	0	LF		0
4″ std pipe, bw, valve, bolt up	0.40	389.0	LF	0	LF		0
6″ std pipe, bw, valve, bolt up	0.67	726.0	LF	0	LF		0
8″ std pipe, bw, valve, bolt up	0.89	227.0	LF	0	LF		0
14″ std pipe, bw, valve, bolt up	1.52	228.0	LF	0	LF		0
18″ std pipe, bw, valve, bolt up	2.03	41.0	LF	0	LF		0
20″ std pipe, bw, valve, bolt up	6.53	74.0	LF	0	LF		0
24′ std pipe, bw, valve, bolt up	2.14	106.0	LF	0	LF		0
30″ std pipe, bw, valve, bolt up	3.13	22.0	LF	0	LF		0
24′ std pipe, bw, valve, bolt up	2.40	3.0	LF	0	LF		0
Hydro test pipe	273.96	1.0	Test	0	Test		0

4.9   Pressure Swing Adsorber (PSA)—Equipment Installation Man Hours

Facility—Refinery and Hydrogen Plant Equipment	Actual	Estimated		
Scope of Work	MH	BM	PF	MH
Equipment erection	380	0		**0**
Offload adsorbers, skids, and surge vessels	340	0		**0**
Set adsorbers V-8000 A, C, E, H, and J (12′ ID°×°45′) 192,400#	580	0		**0**
Install interconnecting skid and vessel piping	2283		0	**0**
Hydro test pipe	274		0	**0**
Equipment/vendor piping installation man hours	**3856**	**0**	**0**	**0**

4.10   Structural Steel Man Hour Table

Facility Class—Refinery and Hydrogen Plant Equipment
    Structural Steel and Miscellaneous Iron
    Erect Structural Steel

	MH/ton
Erect Structural Steel; <=20 tons	
Light—0–19 lb/ft	28.0
Medium—20–39 lb/ft	24.0
Heavy—40–79 lb/ft	20.0
X heavy—80–120 lb/ft	16.0
Erect Structural Steel; 20 tons > tons <= 100 tons	
Light—0–19 lb/ft	21.0
Medium—20–39 lb/ft	18.0
Heavy—40–79 lb/ft	15.0
X heavy—80–120 lb/ft	12.0
Erect Structural Steel; >100 tons	
Light—0–19 lb/ft	16.8
Medium—20–39 lb/ft	14.4
Heavy—40–79 lb/ft	13.2
X heavy—80–120 lb/ft	11.8
	MH/SF
Platform framing	0.15
	MH/LF
Handrail and toe plate	0.25
	MH/SF
Floor grating	0.20
	MH/EA
Stair treads	0.85
Ladders	**MH/LF**
Straight ladder	0.30
Caged ladders	0.35

4.11 WHR Structural Steel Installation Estimate

Facility—Refinery and Hydrogen Plant Equipment		Historical			Estimate		
Description	MH	Quantity	Unit	Quantity	Unit	IW
WHR Structural Steel						
Light—0–19 lb/ft	28.0	42.0	ton	0	ton	0
Medium—20–39 lb/ft	24.0	58.0	ton	0	ton	0
Heavy—40–79 lb/ft	20.0	152.0	ton	0	ton	0
X heavy—80–120 lb/ft	0.0		ton	0	ton	0
Grating	0.20	5552.0	SF	0	SF	0
Handrail and toe plate	0.25	1638.0	LF	0	LF	0
Stair treads	0.85	6.0	EA	0	EA	0
Ladders	0.30	338.0	LF	0	LF	0
Caged ladders	0.35	154.0	LF	0	LF	0

4.12 WHR Structural Steel—Structural Steel Installation Man Hours

Facility—Refinery and Hydrogen Plant Equipment	Actual	Estimated
Scope of Work	MH	IW
Structural Steel Erection Man Hours		
WHR structural steel	**5608**	0
Grating	**1110**	0
Handrail and toe plate	**410**	0
Stair treads	**5**	0
Ladders	**101**	0
Caged ladders	**54**	0
Structural steel erection man hours	**7288**	**0**

4.13 General Scope of Field Work Required for Waste Heat Recovery Equipment

Waste Heat Recover Equipment
  Scope of Work-Field Erection

Install air preheaters
Set reformed gas waste heat boiler
Install steam drum
Set flue gas boiler
Install steam superheaters
Install mix feed superheaters
Install Steam blow down drum
Install steam separators
Erect selective catalytic reduction unit
Module 5, 6, and 7
Install ammonia flow control unit
Install ammonia injection grid
Erect flue gas stack
Install internal piping
Install forced draft fan

4.14 Waste Heat Recovery Equipment Estimate Data

4.14.1 Air Preheaters, Flue Gas Boiler, Steam Superheaters, SCR, Stack Sheet 1

Description	MH	Unit
Install Air Preheaters		
Rig and set combustion air preheaters	0.75	ton
Bolt connection	0.20	EA
Make up anchor bolts	1.00	EA
Set reformed gas waste heat boiler	1.50	ton
Steam drum	2.50	ton
Flue Gas Boiler		
Rig and set flue gas boiler	1.50	ton
Bolt connection	0.20	EA
Make up anchor bolts	1.00	EA
Steam drum	1.50	ton
Install Steam Super Heaters		
Steam super heaters	1.80	ton
Bolt connection	0.20	EA
Make up anchor bolts	0.50	EA
Steam blow down drum	32.00	ton
Steam separators	6.40	ton
Selective catalytic reduction unit (SCR)	6.80	ton
Include: Ammonia Flow Control Unit		
Ammonia Injection Grid		
Flue gas stack	3.36	LF
Forced draft fan	32.00	ton

4.15 Waste Heat Recovery Equipment Installation Estimate

4.15.1 Air Preheaters, Flue Gas Boiler, Steam Superheaters, SCR, Stack Sheet 1

	Historical			Estimate		
Description	MH	Quantity	Unit	Quantity	Unit	BM
Install Air Preheaters						**0**
Rig and set combustion air preheater I	0.75	105.0	ton	0	ton	0
Bolt connection	0.20	112.0	EA	0	EA	0
Make up anchor bolts	1.00	24.0	EA	0	EA	0
Rig and set combustion air preheater II	0.75	105.0	ton	0	ton	0
Bolt connection	0.20	112.0	EA	0	EA	0
Make up anchor bolts	1.00	24.0	EA	0	EA	0
Set Reformed Gas Waste Heat Boiler	1.50	180.0	ton	0	ton	**0**
Steam Drum	2.50	35.0	ton	0	ton	**0**
Flue Gas Boiler						**0**
Rig and set flue gas boiler	1.50	140.0	ton	0	ton	0
Bolt connection	0.20	128.0	EA	0	EA	0
Make up anchor bolts	1.00	24.0	EA	0	EA	0
Steam drum	1.50	140.0	ton	0	ton	0
Install Steam Superheaters						**0**
Steam superheater I	1.80	30.0	ton	0	ton	0
Steam superheater II	1.80	40.0	ton	0	ton	0
Mix feed superheater I	1.80	90.0	ton	0	ton	0
Mix feed superheater II	1.80	60.0	ton	0	ton	0
Bolt connection	0.20	128.0	EA	0	EA	0
Make up anchor bolts	1.00	48.0	EA	0	EA	0
Steam Blow Down Drum	32.00	2.0	ton	0	ton	**0**
Steam Separator	6.40	5.0	ton	0	ton	**0**
Steam Separator	6.40	5.0	ton	0	ton	**0**
Selective Catalytic Reduction Unit (SCR)	6.80	157.1	ton	0	ton	**0**
Includes ammonia flow control unit						
Ammonia injection grid						
Flue Gas Stack	3.36	115.0	LF	0	LF	**0**
Forced Draft Fan	32.00	8.7	ton	0	ton	**0**

4.16 Waste Heat Recovery Equipment—Equipment Installation Man Hours

Facility—Refinery and Hydrogen Plant Equipment	Actual	Estimated
Scope of Work	MH	BM
Install air preheaters	250	0
Set reformed gas waste heat boiler	270	0
Steam drums	88	0
Rig and set flue gas boiler	260	0
Install steam superheaters	470	0
Steam blow down drum	64	0
Steam separators	64	0
Selective catalytic reduction unit (SCR)	1068	0
Flue gas stack	386	0
Forced draft fan	278	0
Equipment installation man hours	3198	0

4.17 Hydrogen Plant Major Equipment Man Hour Breakdown

	Actual	Estimate			
Facility—Hydrogen Plant Equipment	MH	BM	PF	IW	MH
4.5 Stem Methane Reformer	19,538	0			0
4.9 Pressure Swing Adsorber (PSA)	3856	0	0		0
4.12 WHR Structural Steel	7288			0	0
4.16 Waste Heat Recovery Equipment	3198	0			0
Hydrogen Plant Project Man Hours	33,800	0	0	0	0

4.18 General Scope of Field Work Required for Each Electrostatic Precipitator (ESP)

Electrostatic Precipitator (ESP)
Scope of Work-Field Erection
Casing

Prefabricate wall and partition panels
Install anvil beams and antisneakage baffles on end frames/roof beams and bottom baffles
Set partition and wall panels; stabilize with top end frames, bottom end
Frames and intermittently with intermediate roof beams and supporting bottom baffles
Install remaining intermediate roof beams and supporting bottom baffles
Square and plumb casing and began weld out
Complete casing welding, including the hoppers
Drop in preassembled hoppers with lower alignment frames and scaffolding inside
Secure interior cross ties between roof beams, end frames, and bottom baffles
Hang inlet and outlet perforated plates
Preassemble and mount inlet and outlet plenums
Install inlet and outlet plenum cone sections
Install platforms, stairways, and ladders

Internals

Lift and set collecting plates into ESP
Join plate halves and attach plates to anvil beams
Do preliminary alignment of collecting plates
Preassemble discharge electrodes to high voltage frames
Lift high voltage frames and rest on collecting plates

Penthouse

Preassemble hot roof sections
Install hot roof sections
Install lower alignment frames to discharge electrodes
Suspend high voltage frames from support insulators
Complete welding of penthouse walls and hot roof
Weld out rapper sleeves (installed below)

ESP Roof

Preassemble cold roof sections
Insulate underside of cold roof
Set cold roof sections and expansion joints
Install roof handrail
Install and weld out rapper sleeves
Weld out cold roof
Install rapper shafts, rapper adjusting studs, and rappers
Align rappers
Set TR sets
Install duct support steel
Install inlet ductwork on the support steel including the louver dampers
Install seal air system platform on duct support steel
Set seal air fan and heater
Support structure
Sliding plates and cap plates

4.19   Electrostatic Precipitator Equipment Estimate Data

4.19.1   Hoppers, Casing/Partition Frames, Support Steel, Diffusers, TEFs and IRBs Sheet 1

Description	MH	Unit
Preassemble Hoppers		
Place hopper cones and baffles	1.80	ton
Place side wall sections	1.50	ton
Weld side wall and hopper baffles	0.32	LF
Clean out door	4.00	EA
Preassemble Casing and Partition Frames		
Place and fit side and partition frame sections	10.00	ton
Weld side frame and partition sections—3/8″ square bw	0.75	LF
Bevel beams 1/2″ thick × 12″	2.50	EA
Place and fit beams	10.00	EA
Full penetration field weld 1/2″ double bevel	16.80	EA
Erect Precipitator Support Steel		
Erect and bolt structural columns	40.00	EA
Erect X—bracing	21.00	EA
Bolt X—bracing	3.40	EA
Preassemble diffusers	240.00	JOB
Preassemble Perforated Plates		
Load, haul, offload inlet/outlet perforated plates 11′-3″ × 5′-0″	2.00	EA
Place and fit perforated plate sections	4.00	EA
Weld perforate plate sections	0.35	LF
Place and weld clips	1.00	EA
Preassemble TEFs and IRBs		
Load, haul, offload frames	2.76	ton
Set top end frames	20.00	EA
Set intermediate roof beams	32.00	EA

4.19.2   Slide Plates, Hopper Platform, Support Steel Access Sheet 2

Description	MH	Unit
Set Precipitator Slide Plates		
Set slide plates	4.00	EA
Erect Hopper Platform		
Erect platform steel	0.10	SF
Erect grating	0.15	SF
Erect handrail/toe plate	0.25	LF
Install Support Steel Access		
Erect platform steel	0.29	SF
Erect grating	0.24	SF
Erect handrail/toe plate	0.36	LF
Stairs	0.85	LF
Ladder	0.30	LF

4.19.3 Penthouse, Casing, Hoppers, Weld Casing, Frames/Beams Sheet 3

Description	MH	Unit
Preassemble Penthouse		
Place and fit penthouse roof and side frame sections	2.50	ton
Weld penthouse roof sections and sides/ends	2.15	LF
Erect Casing		
Rig and place side and partition frame casing	40.00	EA
Attach temporary guy lines to casing/turn buckles	6.67	EA
Set Hoppers		
Rig and place hoppers on support steel	1.50	ton
Field bolt list	0.15	EA
Plumb, Square, Weld Casing		
Plumb, square, and fit casing (turnbuckle)	360.00	JOB
Weld side frame and partition sections—3/8″ square bw	0.75	LF
Weld side wall casing corners	0.35	LF
Weld outside side wall casing 2-10	0.35	LF
Bevel Beams 1/2″ thick × 12″	2.50	EA
Place and fit beams	10.00	EA
Full penetration field weld 1/2″ double bevel	16.00	EA
Field bolts	300.00	Lot
Install Top End Frames and Intermediate Roof Beams		
Place and fit top end frames/roof beams	40.00	ton
Field bolts	0.20	EA
Seal weld around bolts/field weld 1/4″ fillet	0.35	LF
Full penetration field weld 1/2″ double bevel	16.00	EA

4.19.4 Frames/Baffles, Collecting Plates, Hopper Auxiliaries, Electrodes, Hot Roof Sheet 4

Description	MH	Unit
Install Bottom End Frames and Bottom Support Baffles		
Place and fit bottom end frames, and baffles	2.50	ton
Field bolts	0.20	EA
Seal weld around bolts/field weld 1/4″ fillet	0.35	LF
Full penetration field weld 1/2″ double bevel	16.00	EA
Install Collecting Plates		
Install collecting plates	3.50	EA
Bolt plates	1.00	EA
Weld corner	240.00	JOB
Decking	160.00	JOB
Install Hopper Auxiliaries		
Install pneumatic hopper slide gates (12″ × 12″)	7.80	Assembly
Install manual hopper slide gates (12″ × 12″)	7.80	Assembly
Install TR controllers	6.00	Assembly
Install Upper Frames/Electrodes		
Install high voltage frames	16.00	EA
Load, haul, offload discharge electrodes	0.50	EA
Install spiked discharge electrodes 2″ diameter × 40′	1.20	EA
Preliminary Alignment	480.00	JOB
Install/Weld Hot Roof		
Place and fit casing roof sections	27.50	ton
Weld roof casing	0.35	LF
Weld roof casing at corners	0.35	LF

4.19.5 Set Plates, Rapper Rods, Manufacturing Plates, Penthouse, Diffuser, Alignment, Rectifiers Sheet 5

Description	MH	Unit
Set Precipitator Perforated Plates		
Set perforated plate sections	10.00	EA
Weld perforated plate sections	0.35	LF
Place and weld clips	1.50	EA
Set Rapper Rods		
Rapper sleeves, shafts and insulators, DE alumina rapper shafts	1.10	EA
Seal weld sleeves	0.35	EA
Set Support Bushings/Manufacturing Plates		
Set and weld, bolt manufacturing plates	1.00	EA
Set support bushings	0.24	EA
Erect Penthouse		
Place and fit penthouse roof sections, sides and ends	2.5	ton
Weld penthouse roof, sides, ends, and corners	2.15	LF
Erect and weld in/out diffusers	1.00	EA
Final alignment	2.10	ton
Set Transformer Rectifier Sets		
Load, haul, offload transformer rectifiers	4.00	EA
Set Transformer rectifiers	20.00	EA
Install rappers	3.00	EA
Install rapper controllers	20.00	EA

4.19.6 Seal Air System, Dampers, Stack, Inlet/Outlet Plenum Sheet 6

Description	MH	Unit
Erect Seal Air System		
Install seal air blower and heater system	260.00	EA
Install Dampers and Access Doors		
Install dampers	50.00	EA
Install 30″ × 30″ access doors	4.00	EA
Erect Stack		
Set bottom and upper section	60.00	EA
Bolt bottom section to foundation	1.53	EA
Fit and weld upper sections	0.75	LF
Platform—360 degrees	0.10	SF
Grating	0.15	SF
Handrail	0.25	LF
Step platforms	6.00	EA
Caged ladder	0.35	LF
Preassemble and Erect Inlet/Outlet Plenum		
Place and fit inlet and outlet plenum sections	11.40	ton
Weld inlet and outlet plenum sections	0.75	LF
Erect Inlet/Outlet Plenum		
Place and fit inlet plenum sections	22.00	ton
Place and fit outlet plenum sections	40.00	ton
Weld inlet and outlet plenum sections	0.75	LF

4.19.7 Penthouse Structural Steel Sheet 7

Description	MH	Unit
Penthouse Structural Steel and Platforms		
Erect structural steel	8.00	pcs
Bolted connections	1.00	EA
Platforms and stairs	160.00	Lot

4.20 Electrostatic Precipitator Equipment Installation Estimate

4.20.1 Hoppers, Casing/Partition Frames, Erect Steel, Diffusers, TEF/IRB Sheet 1

		Historical		Estimate		
Description	MH	Quantity	Unit	Quantity	Unit	BM
Pre Assemble Hoppers						0
Set hopper cones 20 EA	1.80	26.4	ton	0	ton	0
Place and fit hopper side wall sections	1.50	125.1	ton	0	ton	0
Weld side wall sections	0.32	3600.0	LF	0	LF	0
Place and fit hopper baffle	1.80	11.1	ton	0	ton	0
Weld hopper Baffle 3/16″ fillet/1/8″ fillet 2-10	0.32	1520.0	LF	0	LF	0
Clean out door	4.00	20.0	EA	0	EA	0
Preassemble Casing and Partition Frames						0
Place and fit side frame sections	10.00	119.4	ton	0	ton	0
Weld and fit side frame sections—3/8″ square bw	0.75	364.0	LF	0	LF	0
Place and fit partition frame sections	10.00	35.8	ton	0	ton	0
Weld and fit partition frame sections—3/8″ square bw	0.75	182.0	LF	0	LF	0
Bevel beams 1/2″ thick × 12″	2.50	56.0	EA	0	EA	0
Place and fit beams	10.00	32.0	EA	0	EA	0
Full penetration field weld 1/2″ double bevel	16.80	56.0	EA	0	EA	0
Erect Precipitator Support Steel						0
Erect and bolt structural columns	40.00	21.0	EA	0	EA	0
Erect X—bracing	21.00	22.0	EA	0	EA	0
Bolt X—bracing	3.40	104.0	EA	0	EA	0
Preassemble Diffusers	240.00	1.0	JOB	0	JOB	0
Preassemble Perforated Plates						0
Load, haul, offload inlet/outlet perforated plates	2.00	56.0	EA	0	EA	0
Place and fit perforated plate sections	4.00	56.0	EA	0	EA	0
Weld perforated plate sections	0.35	576.0	LF	0	LF	0
Place and weld clips	1.00	40.0	EA	0	EA	0
Preassemble TEFs and IRBs						0
Load, haul, offload frames	2.76	37.5	ton	0	ton	0
Set top end frames	20.00	4.0	EA	0	EA	0
Set intermediate roof beams	32.00	8.0	EA	0	EA	0

4.20.2   Slide Plates, Hopper Platform, Support Steel Access Sheet 2

Description	MH	Historical Quantity	Unit	Estimate Quantity	Unit	BM
Set Precipitator Slide Plates						**0**
Set slide plates	4.00	21.0	EA	0	EA	0
Erect Hopper Platform						**0**
Erect platform steel	0.10	7700.0	SF	0	SF	0
Erect grating	0.15	7700.0	SF	0	SF	0
Erect handrail	0.25	360.0	LF	0	LF	0
Install Support Steel Access						**0**
Stair towers from side access level to top ESP roof						
Erect platform steel	0.29	89.4	SF	0	SF	0
Erect grating	0.24	128.0	SF	0	SF	0
Erect handrail/toe plate	0.36	288.0	LF	0	LF	0
Stairs	0.85	16.0	EA	0	EA	0
Stair towers from grade to side access level						
Erect platform steel	0.29	128.0	SF	0	SF	0
Erect grating	0.24	128.0	SF	0	SF	0
Erect handrail/toe plate	0.36	208.0	LF	0	LF	0
Stairs	0.85	16.0	EA	0	EA	0
Side access level				0		
Erect platform steel	0.29	592.0	SF	0	SF	0
Erect grating	0.24	592.0	SF	0	SF	0
Erect handrail/toe plate	0.36	188.0	LF	0	LF	0
Ladder	0.30	16.0	LF	0	LF	0
Caged ladders from grade to roof						
Erect platform steel	0.29	160.0	SF	0	SF	0
Erect grating	0.24	160.0	SF	0	SF	0
Erect handrail/toe plate	0.36	120.0	LF	0	LF	0
Ladder	0.30	200.0	LF	0	LF	0
Side access to hopper platform						
Erect platform steel	0.29	64.0	SF	0	SF	0
Erect grating	0.24	64.0	SF	0	SF	0
Erect handrail/toe plate	0.36	48.0	LF	0	LF	0
Handrail on ESP roof perimeter						
Erect handrail/toe plate	0.36	360.0	LF	0	LF	0

4.20.3   Penthouse, Casing, Hoppers, Weld Casing, Frames/Beams Sheet 3

Description	Historical			Estimate		
	MH	Quantity	Unit	Quantity	Unit	BM
Preassemble Penthouse						**0**
Place and fit penthouse roof sections	2.50	31.6	ton	0	ton	0
Weld penthouse roof sections	2.15	280	LF	0	LF	0
Place and fit penthouse sides and ends	2.50	18.4	ton	0	ton	0
Weld penthouse sides and ends	2.15	200	LF	0	LF	0
Erect Casing						**0**
Rig and place side frame casing (rows 1 and 3)	40.00	6.00	EA	0	EA	0
Rig and place partition frame casing (row 2)	40.00	3.00	EA	0	EA	0
Attach temporary guy lines to casing/turn buckles	6.67	36.00	EA	0	EA	0
Set Hoppers						**0**
Rig and place hoppers on support steel	1.50	162.3	ton	0	ton	0
Field bolt list	0.15	2294.0	EA	0	EA	0
Plumb, Square, Weld Casing						**0**
Plumb, square, and fit casing (turnbuckle)	360.00	1.0	JOB	0	JOB	0
Weld side frame sections—3/8″ square bw	0.75	200.0	LF	0	LF	0
Weld partition frame sections—3/8″ square bw	0.75	100.0	LF	0	LF	0
Weld side wall casing corners	0.35	500.0	LF	0	LF	0
Weld outside side wall casing 2-10	0.35	540.0	LF	0	LF	0
Bevel beams 1/2″ thick × 12″	2.50	80.0	EA	0	EA	0
Place and fit beams	10.00	44.0	EA	0	EA	0
Full penetration field weld 1/2″ double bevel	16.00	80.0	EA	0	EA	0
Field bolts	300.00	1.0	Lot	0	Lot	0
Install Top End Frames and Intermediate Roof Beams						**0**
Place and fit top end frames	40.00	16.3	ton	0	ton	0
Place and fit roof beams	40.00	32.4	ton	0	ton	0
Field bolts	0.20	144.0	EA	0	EA	0
Seal weld around bolts	0.35	144.0	EA	0	EA	0
Field weld 1/4″ fillet	0.35	200.0	LF	0	LF	0
Full penetration field weld 1/2″ double bevel	16.00	24.0	EA	0	EA	0

4.20.4 Frames/Baffles, Collecting Plates, Hopper Auxiliaries, Electrodes, Hot Roof Sheet 4

Description	Historical			Estimate		
	MH	Quantity	Unit	Quantity	Unit	BM
Install Bottom End Frames						**0**
and Bottom Support						
Baffles						
Place and fit bottom end frames	2.50	14.3	ton	0	ton	0
Place and fit baffles	2.50	28.6	ton	0	ton	0
Field bolts	0.20	144.0	EA	0	EA	0
Seal weld around bolts	0.35	144.0	LF	0	LF	0
Field weld 1/4″ fillet	0.35	200.0	LF	0	LF	0
Full penetration field weld 1/2″ double bevel	16.00	24.0	EA	0	EA	0
Install Collecting Plates						**0**
Install collecting plates	3.50	700.0	EA	0	EA	0
Bolt plates	1.00	350.0	EA	0	EA	0
Weld corner	240.00	1.0	JOB	0	JOB	0
Decking	160.00	1.0	JOB	0	JOB	0
Install Hopper Auxiliaries						**0**
Install pneumatic hopper slide gates (12″×12″)	7.80	20.0	Assembly	0	Assembly	0
Install manual hopper slide gates (12″×12″)	7.80	20.0	Assembly	0	Assembly	0
Install TR controllers	6.00	20.0	Assembly	0	Assembly	0
Install Upper Frames/						**0**
Electrodes						
Install high voltage frames	16.00	40.0	EA	0	EA	0
Load, haul, offload discharge electrodes	0.50	2720.0	EA	0	EA	0
Install spiked discharge electrodes 2″ diameter	1.20	2720.0	EA	0	EA	0
Preliminary alignment	480.00	1.0	JOB	0	JOB	0
Install/Weld Hot Roof						**0**
Place and fit casing roof sections	27.50	55.0	ton	0	ton	0
Weld roof casing	0.35	1660.0	LF	0	LF	0
Weld roof casing at corners	0.35	720.0	LF	0	LF	0

4.20.5   Perforated Plates, Rapper Rods, Manufacturing Plates, Penthouse, Diffuser, Alignment, Rectifier Set Sheet 5

Description	Historical			Estimate		
	MH	Quantity	Unit	Quantity	Unit	BM
Set Precipitator Perforated Plates						**0**
Set perforated plate sections	4.00	16.0	EA	0	EA	0
Weld perforated plate sections	0.35	384.0	LF	0	LF	0
Place and weld clips	1.00	8.0	EA	0	EA	0
Set Rapper Rods						**0**
Set plate rapper sleeves 4″ diameter × 10′	1.10	340.0	EA	0	EA	0
Seal weld sleeves	0.35	340.0	EA	0	EA	0
Set plate rapper shafts 2″ diameter × 10′	1.10	340.0	EA	0	EA	0
Set alumina support insulators	1.10	80.0	EA	0	EA	0
Set DE alumina rapper shafts	1.10	80.0	EA	0	EA	0
Set Support Bushings/ Manufacturing Plates						**0**
Set manufacturing plates	1.00	340.0	EA	0	EA	0
Bolt and weld manufacturing plates	1.00	340.0	EA	0	EA	0
Set support bushings	0.24	340.0	EA	0	EA	0
Erect Penthouse						**0**
Place and fit penthouse roof sections	2.5	31.6	ton	0	ton	0
Weld penthouse roof sections	2.15	1140.0	LF	0	LF	0
Place and fit penthouse sides and ends	2.5	18.4	ton	0	ton	0
Weld penthouse sides and ends	2.15	160.0	LF	0	LF	0
Weld penthouse sides and ends at corners	2.15	700.0	LF	0	LF	0
Erect and Weld In/Out Diffusers	1.00	600	EA	0	EA	**0**
Final Alignment	2.10	1986.5	ton	0	ton	**0**
Set Transformer Rectifier Sets						**0**
Load, haul, offload transformer rectifiers	4.00	20.0	EA	0	EA	0
Set transformer rectifiers	20.00	20.0	EA	0	EA	0
Install Rappers						**0**
Install rappers	3.00	426.0	EA	0	EA	0
Install rapper controllers	20.00	2.0	EA	0	EA	0

4.20.6 Seal Air System, Dampers, Stack, and Inlet/Outlet Plenum Sheet 6

	Historical			Estimate		
Description	MH	Quantity	Unit	Quantity	Unit	BM
Erect Seal Air System						**0**
Install seal air blower and heater system	260.00	2.0	EA	0	EA	0
Install Dampers and Access Doors						**0**
Install dampers	50.00	2.0	EA	0	EA	0
Install 30″×30″ access doors	4.00	44.0	EA	0	EA	0
Erect Stack						**0**
Set bottom section	60.00	1.0	EA	0	EA	0
Bolt bottom section to foundation	1.53	28.0	EA	0	EA	0
Set upper sections	60.00	3.0	EA	0	EA	0
Fit and weld upper sections	0.75	226.2	LF	0	LF	0
Install platforms at ground level						
Platform—360 degrees	0.10	527.8	SF	0	SF	0
Grating	0.15	527.8	SF	0	SF	0
Handrail	0.25	150.0	LF	0	LF	0
Step platforms	6.00	4.0	EA	0	EA	0
Caged ladder	0.35	140.0	LF	0	LF	0
Preassemble Inlet/Outlet Plenum						**0**
Place and fit inlet plenum sections	11.40	63.9	ton	0	ton	0
Weld inlet plenum sections	0.75	1240.0	LF	0	LF	0
Place and fit outlet plenum sections	11.40	91.1	ton	0	ton	0
Weld outlet plenum sections	0.75	1340.0	LF	0	LF	0
Erect Inlet/Outlet Plenum						**0**
Place and fit inlet plenum sections	22.00	40.8	ton	0	ton	0
Weld inlet plenum sections	0.75	800.0	LF	0	LF	0
Place and fit outlet plenum sections	40.00	114.2	ton	0	ton	0
Weld outlet plenum sections	0.75	1360.0	LF	0	LF	0

4.20.7 Penthouse Structural Steel Sheet 7

	Historical			Estimate		
Description	MH	Quantity	Unit	Quantity	Unit	BM
Penthouse Structural Steel and Platforms						**0**
Erect structural steel	8.00	21.0	pcs	0	pcs	0
Bolted connections	1.00	108.0	EA	0	EA	0
Platforms and stairs	160.00	1.0	Lot	0	Lot	0

4.21 Electrostatic Precipitator Equipment—Equipment Installation Man Hours

Facility—Refinery and Hydrogen Plant Equipment	Actual	Estimated
Scope of Work	MH	BM
Preassemble hoppers	1974	0
Pre assemble casing and partition frames	3362	0
Erect precipitator support steel	1656	0
Preassemble diffusers	240	0
Preassemble perforated plates	578	0
Preassemble TEFs and IRBs	440	0
Set precipitator slide plates	84	0
Erect hopper platform	2025	0
Install support steel access	1082	0
Preassemble penthouse	1157	0
Erect casing	600	0
Set hoppers	588	0
Plumb, square, weld casing	3169	0
Install top end frames and intermediate roof beams	2481	0
Install bottom end frames and bottom support baffles	640	0
Install collecting plates	3200	0
Install hopper auxiliaries	432	0
Install upper frames/electrodes	5264	0
Preliminary alignment	480	0
Install/weld hot roof	2346	0
Set precipitator perforated plates	206	0
Set rapper rods	1043	0
Set support bushings/Manufacturing plates	760	0
Erect penthouse	4425	0
Erect and weld in/out diffusers	600	0
Final alignment	4172	0
Set transformer rectifier sets	480	0
Install rappers	1320	0
Erect seal air system	520	0
Install dampers and access doors	276	0
Erect stack	696	0
Preassemble inlet/outlet plenum	3702	0
Erect inlet/outlet plenum	7086	0
Penthouse structural steel and platforms	436	0
Equipment installation man hours	**57,518**	**0**

Compressor Station Equipment

5.1 Compressor Station Estimate Data

5.1.1 Foundation, Driver, Motor, Gearbox and Compressor, Loose Components, Connections Sheet 1

Description	MH	Unit
Prep foundation	120.00	Job
Driver packages, motor drive, gearbox, and compressor	4.76	ton
Final alignment	4.76	ton
Install jacking bolts	8.00	EA
Grout skid	1.40	SF
Compressor flanges—bolted connection 20″ 900#	11.00	EA
Compressor flanges—bolted connection 1-1/2″ 900#	1.30	EA
Loose Components		
Lube oil mist separator	20.0	EA
Flame arrestor, lube oil tank vent	20.0	EA
Flame arrestor, dry seal vent, primary/secondary	40.0	EA
Filter strainer	20.0	EA
Unit control console, 1 Bay	80.0	EA
Lube oil cooler	100.0	EA
VFD lube oil pump	20.0	EA
Connections		
4″ 150# lube oil tank vent	2.6	EA
2″ 150# pneumatic postlube backup vent	1.8	EA
2″ cap lube oil tank fill	1.8	EA
3/4″ lube oil cooler vent	1.8	EA
3″ 150# lube oil	3.6	EA
1″ 150# filter/lube oil	5.4	EA
1″ 1500# filter/lube oil	3.6	EA
2″ NPT lube oil	1.8	EA
2″ 300# pneumatic postlube backup supply	1.8	EA
2″ NPT lube oil pump motor	3.6	EA
6″ 150# oil mist separator	7.8	EA
6″ 125# flame arrestor	3.9	EA
2″ 125# flame arrestor	1.8	EA
2″ 300# strainer	3.6	EA
3″ 150# lube oil cooler	3.9	EA
Miscellaneous	10.8	EA

Industrial Piping and Equipment Estimating Manual. http://dx.doi.org/10.1016/B978-0-12-813946-2.00005-8

5.2 Compressor Station Installation Estimate

5.2.1 Foundation, Driver, Motor, Gearbox and Compressor, Loose Components, Connections Sheet 1

Description	MH	Historical Quantity	Unit	Estimate Quantity	Unit	BM	MW	PF
Prep foundation	120.0	1.0	Job	0	Job		0	
Driver packages, motor drive,						0		
gearbox, and compressor								
Driver package	4.76	50.4	ton	0	ton	0		
Driven package	4.76	20.3	ton	0	ton	0		
Electric motor drive	4.76	20.7	ton	0	ton	0		
Gearbox	4.76	12.9	ton	0	ton	0		
Compressor C402	4.76	13.5	ton	0	ton	0		
Alignment skid	4.76	54.6	ton	0	ton		0	
Install jacking bolts	8.0	20.0	EA	0	EA		0	
Grout skid	1.4	200.0	SF	0	SF		0	
Compressor flanges—bolted								0
connection								
Suction 20″ 900#	11.0	1.0	EA	0	EA			0
Discharge 20″ 900#	11.0	1.0	EA	0	EA			0
Vent and drain flanges 1-1/2″	1.3	24.0	EA	0	EA			0
900#								
Loose components						0		
Lube oil mist separator	20.0	1.0	EA	0	EA	0		
Flame arrestor, lube oil tank vent	20.0	1.0	EA	0	EA	0		
Flame arrestor, dry seal vent,	40.0	2.0	EA	0	EA	0		
primary/secondary								
Filter strainer	20.0	1.0	EA	0	EA	0		
Unit control console, 1 Bay	80.0	1.0	EA	0	EA	0		
Lube oil cooler	100.0	1.0	EA	0	EA	0		
VFD lube oil pump	20.0	1.0	EA	0	EA	0		
Connections								0
4″ 150# Lube oil tank vent	2.6	1.0	EA	0	EA			0
2″ 150# Pneumatic postlube	1.8	1.0	EA	0	EA			0
backup vent								
2″ Cap lube oil tank fill	1.8	1.0	EA	0	EA			0
3/4″ Lube oil cooler vent	1.8	1.0	EA	0	EA			0
3″ 150# Lube oil	3.6	2.0	EA	0	EA			0
1″ 150# Filter/lube oil	5.4	3.0	EA	0	EA			0
1″ 1500# Filter/lube oil	3.6	2.0	EA	0	EA			0
2″ NPT lube oil	1.8	1.0	EA	0	EA			0
2″ 300# Pneumatic postlube	1.8	1.0	EA	0	EA			0
backup supply								
2″ NPT lube oil pump motor	3.6	2.0	EA	0	EA			0
6″ 150# Oil mist separator	7.8	2.0	EA	0	EA			0
6″ 125# Flame arrestor	3.9	1.0	EA	0	EA			0
2″ 125# Flame arrestor	1.8	1.0	EA	0	EA			0
2″ 300# Strainer	3.6	2.0	EA	0	EA			0
3″ 150# Lube oil cooler	3.9	2.0	EA	0	EA			0
Miscellaneous	10.8	6.0	EA	0	EA			0

5.3 Compressor Station–Equipment Installation Man Hours

Facility Compressor Station	Actual		Estimated		
	MH	BM	MW	PF	MH
Prep foundation	120		0		0
Driver packages, motor drive, gearbox, and compressor	560	0			0
Alignment skid	260		0		0
Install jacking bolts	160		0		0
Grout skid	280		0		0
Compressor flanges—bolted connection	53			0	0
Loose components	340	0			0
Connections	151			0	0
Equipment and piping installation man hours	**1924**	0	0	0	0

5.4 Cooling System Estimate Data

5.4.1 Duct Work, Bolted Joints, and Duct Support Sheet 1

Description	MH	Unit
Install Duct Work		
Flue, duct, hood, including supports, hoppers, expansion joints, and dampers	48.00	ton
Filter	28.00	EA
Field weld (5/16″ square joint)	0.75	LF
Silencer	30.00	EA
Fan	40.00	EA
Flex section	2.00	LF
Bolt connection—field bolt up	0.25	EA
Duct support	2.00	EA
Base plate AB	2.00	EA
Base plate grout	4.00	EA
Support columns and beams	6.00	EA
Duct support clips	1.01	EA
2×2 Angle	3.00	EA
Receive, store, and haul to hook	600.00	Job

5.5 Compressor Equipment Installation Estimate

5.5.1 Duct Work, Bolted Joints, and Duct Support Sheet 1

	Historical			Estimate		
Description	MH	Quantity	Unit	Quantity	Unit	BM
Install duct work						**0**
Weld joint	0.75	506.0	LF	0	LF	0
Duct (set and fit up)	48.00	1.5	ton	0	ton	0
Duct ell (set and fit up)	48.00	1.5	ton	0	ton	0
Transition (set and fit up)	48.00	0.6	ton	0	ton	0
Hood (set and bolt down)	48.00	0.4	ton	0	ton	0
Duct wye (set and fit up)	48.00	0.2	ton	0	ton	0
Duct tee (set and fit up)	48.00	0.3	ton	0	ton	0
Damper (set and fit up)	48.00	4.8	ton	0	ton	0
Hartzell fan (set and fit up)	40.00	2.0	EA	0	EA	0
Silencer (set and fit up)	30.00	2.0	EA	0	EA	0
Flex section (set and fit up)	2.00	6.0	LF	0	LF	0
Filter (set and fit up)	28.00	1.0	EA	0	EA	0
Bolted joints						**0**
Hartzell fan 65–35	0.25	20.0	EA	0	EA	0
Exciter elbow	0.25	8.0	EA	0	EA	0
Silencer and duct flanges	0.25	32.0	EA	0	EA	0
Damper D-4, silencer, elbow, flex joint, and transition T-1	0.25	114.0	EA	0	EA	0
Elbow flanges	0.25	72.0	EA	0	EA	0
Flex joint F-2 and F-3, transition T-2 and T-3	0.25	80.0	EA	0	EA	0
Damper D-3	0.25	24.0	EA	0	EA	0
Flex joint F-4, transition T-4	0.25	16.0	EA	0	EA	0
Receive, store, and haul to hook	600.00	1.0	Job	0	Job	**0**
Duct support						**0**
Support columns—vertical	6.00	10.0	EA	0	EA	0
Base plate AB	2.00	32.0	EA	0	EA	0
Base plate grout	4.00	10.0	EA	0	EA	0
Horizontal beams	6.00	8.0	EA	0	EA	0
Field weld horizontal beam to columns	0.75	16.0	LF	0	LF	0
Duct support clips	1.01	32.0	EA	0	EA	0
2×2 angle	3.00	16.0	EA	0	EA	0
Field weld @ angle	2.00	32.0	LF	0	LF	0

5.6 Cooling System–Equipment Installation Man Hours

	Actual	Estimate
Facility Compressor Station	**MH**	**BM**
Install duct work	1000	0
Bolted joints	92	0
Duct support	368	0
Receive, store, and haul to hook	600	0
Equipment installation man hours	**2060**	**0**

Biomass Plant Equipment

6.1 Equipment Descriptions

6.1.1 Equipment Descriptions— Boiler and Fuel Feeding Equipment

The project includes the installation of a new biomass boiler and the installation of a new fuel-handling facility. Erection of the boiler, stoker, economizer, precipitator, mechanical collector, fans, air heater, and ductwork for the boiler island.

Boiler-Pressure Parts

The furnace tubes and generating bank tubes will be shipped knocked down with as much factory assembly as practical for shipping.

Steam Drum and Mud Drum

Install one steam drum and one mud drum. The steam drum will come with all internals mounted. Remove and reinstall after boil out.

Furnace Panels

The furnace consists of wall panels and roof panels.

Water Wall Panels

Install water wall panels.

Coen 100 MMBTU Burner With Primary Air Fan

Set and seal weld the burner wind box to the casing on the furnace. Set IC ductwork from the wind box to the new fan including expansion joint and silencer. New fan will be mounted on the structural steel. Set all loose spool piping assemblies including the jackshaft assemblies. Fabricate and install silencer support.

Install Super Heater Section

Connection to Drums

Generating bank, boiler, sidewall, furnace, roof and furnace rear wall tubes are expanded and flared to 1800 psi.

Soot Blowers

Install rotary and retractable soot blowers and support bearings.

Fuel Feeding Equipment

Metering Bins

Install six metering bin screws. Installation includes support steel. Metering bins will set on load cells. Install 24 load cells.

Boiler Drag Chain

Install one boiler drag chain above the metering bins.

Install 8'-0" × 6'-0" enclosure and six rack and pinions, slide gates, including chain wheel actuators between the meters.

Solid fuel is fed to the boiler in six feed points, all on the front wall.

Fuel Chutes

Install metering bin discharge chutes, expansion joints to stoker windswept spouts, support bins, and drag chains for all six bins.

Industrial Piping and Equipment Estimating Manual. http://dx.doi.org/10.1016/B978-0-12-813946-2.00006-X

6.1.2   Equipment Descriptions—Stoker

Vibrating hydro grate arranged for front ash discharge.

Vibrating Grate Modules

Four shop-assembled grate modules, complete with support steel to grate. Each module will be complete with the following components:

One set shop-installed grate.
Four H-Type "K" thermocouple assemblies with clamps, fitting, IC wire to junction box.
One stationary frame for support of the vibrating frame.
One set of structural legs.
One set flexing straps for support of the vibrating grid section.

Grate Drive Arrangement

One drive arrangement, per stoker module consisting of one 7.5 HP motor, belt drive, drive guard, pillow block bearings, and drive shaft with eccentric linked to vibrating frame.

Interior Stoker Seals

One set "T" seals between stationary and vibrating sections of the stoker.

Seals for Settings

One set high-pressure air seals at the sides, front, and rear of the setting. Seals attach to the stationary frames and the boiler side and rear wall headers. Install and vulcanize expansion Joint.

Distribution Spout Air System

One set high-pressure air ducting with bracing, including manual dampers to serve refuse distributor spouts.

Fuel Distributor Mounting Plate

Reinforced steel mounting plates to support the fuel distributors.

Extension Front Arrangement

Extension front is designed with removable steel front panels for access to the boiler. Front panels, side panels, and roof panels for field installation.

Stoker Doors

Access panels in sides of stationary frame, for floor-set installations.

Refractory-lined hinged access doors with observation ports for each grate module mounted in extension front panels.

Fuel Distributors

Six spouts located on distributor mounting plates. Spouts are $36'' \times 10''$ with high-pressure adjustable air-swept floor for control.

Rotating Air Dampers

Six rotating air damper assemblies to be mounted on the air inlet of the refuse distributor spouts. Dampers are driven by one 1/3 HP gear motor through interconnecting shafting.

Balanced Dampers

Six balanced damper assemblies mounted to the inlet of the wood distributor spouts.

Overfire Air Turbulence System

One Detroit Stoker Company overfire turbulence system, consisting of eight high-pressure overfire air heaters and nozzles, including one manual control damper at one end of each row of header.

6.1.3   Equipment Descriptions—Back Pass Equipment

Economizer
Set one modular economizer complete with access platforms and inlet/outlet duct. Install support steel and platforms.

Air Heaters
Set two ECO tubular air heaters with interconnecting ductwork.

Mechanical Collector
One Warren Environmental Model 288-12-A unit with four hoppers.
Install collector with support steel and access platforms.

CO and No$_x$ Removal System
One model C1 × 192(H) CO and one model N2 × 256 (H) No$_x$ removal catalyst system
CO catalyst removal system will have the following design features:

Catalyst layers 1
CO catalyst layer depth (inches) 2.0

NO$_x$ catalyst removal system will have the following design features:

No$_x$ reduction reagent anhydrous ammonia
Flow distribution devices as required
Catalyst layers 2
Ammonia injection grid 1
Ammonia injection grid with adjustable lances and distribution headers with nozzles
Catalyst will be mounted in honeycomb blocks—18″ × 22.5″
Catalyst housing shell will be seal welded to form a gas tight structure
Shell—25′-9″ H × 25′-0″ W × 45′-0″ H

Wet Scrubber

Amerair Standard Horizontal Quencher. Six spray headers mounted on nozzles.
Amerair Standard Packed Tower—22′-0″ diameter by 39′ straight side. Unit shipped broken down in half—moon sections—12′ W × 24′ L × 12′ H. All sections are flanged for field bolting. Seal seams on inside only; apply 2″ wide fiberglass matte strips followed by a 3″ wide Nexus veil strip coated with fiberglass gel.
There is a 6′-0″ tall bed of polypropylene randomly dumped packing.
Field-installed mist eliminator modules held down by FRP Beams.

Stack
One filament wound FRP Stack 10′ I.D. × 128′ H with open bottom with flange, open top with 10′ × 9′-4″ reducer. Stack has following features:

Four FRP body flanges
Three carbon steel galvanized guy/brace points (4) clips each at 24′ elevation from bottom of stack
Six carbon steel galvanized lifting lugs
Sixty FRP lightening protection clips

6.1.4   Equipment Descriptions—Precipitator

Precipitator

One PPC Model 59R-1534-5712P modular electrostatic precipitator including all collecting plates, rigid discharge electrodes, roof sections, insulator compartments, access doors, and all internal components.

Electrostatic precipitator will have the following design features:

Collecting area (SF)	204,779 SF
Design temperature	700°F
Design pressure	±35
Hopper capacity (CF)	13,045 CU FT
Number of hoppers	5 EA
Hopper opening size	$18' \times 54'\text{-}0''$
Number of gas passages (inches)	59
Spacing of gas passages (inches)	12
Total length of discharge electrodes	70,210 LF
Number of transformers	5 EA
Installed weight (excluding dust, insulation, and support steel)	910,600 lb

The following components will be provided:

Collecting plates—both top and bottom alignment guides, stiffeners, and mountings will maintain alignment of the collecting plates
Electromagnetic uplift-gravity impact rappers
Discharge electrode frames
Step-up transformers/rectifiers
High tension support insulators
Welded weatherproof individual compartments
Precipitator shell will be fabricated from 3/16″ thick A 36 steel plate. Stiffeners to support as required. Shell seal welded for a totally gas tight structure.
Transverse trough-type hoppers
Precipitator supports: All structural steel slide plates
Nozzles; flanged inlet and outlet nozzles
Roof equipment enclosure—mounted on roof
Purge blower control panel

6.1.5   Equipment Descriptions—Ductwork and Breeching and Bottom Ash Drag System

Ductwork and Breeching

Install ductwork and breeching and expansion joints. Fit-up and seal welding from inside. Fit-up flanges tack welded to sections to install and bolt field joints.

Ductwork will rest on sliding supports or hang from supports.

Bottom Ash Drag System

Equipment consists of refractory-lined bottom ash chute and one submerged ash drag chain conveyor.

Bottom ash chute is 3/8″ carbon steel, 34′-0″ L×3′-0″ W×1′-0″ H.

Submerged ash drag conveyor is 3/8″ carbon steel×66′-0″ L with outlet chute.

Shipped in sections:

Head section—motor and drive shipped attached to head
Neck section
Transition section
Straight section
Tail section-take-up shipped assembled

Scope of work:

Assembly of housing
Assembly of drag chain
Installation of drive units
Pneumatic/hydraulic take-up units, including air piping
Discharge chutes
Grouting
Refractory-lined chute on inlet of submerged ash drag
All water piping to and from submerged ash drag and drains

6.1.6   Equipment Descriptions—Fan System

Fans

Perform all millwright work for the four fans; alignment and grouting.

ID Fan, OFA Fan, and FD Fan

Install one Process Barron 2000 HP ID Fan
Install one Process Barron 800 HP OFA Fan
Install one Process Barron 400 HP FD Fan

Work scope:

Set housing and inlet box
Set shaft and inlet cones
Install motor sole plates
Set inlet damper and damper drive
Set motor
Set inlet silencer
Lubrication
Laser alignment
Final dowelling
Grouting of bearings and motor

Spout Air Fan

Install one industrial air products 200 HP spout air fan
Set unitary base housing, motor, and inlet box
Set shaft and inlet cones
Set two ambient air inlet dampers and damper drive
Set FGR inlet damper and damper drive
Set inlet silencer
Lubrication
Laser alignment
Final dowelling
Grouting of bearings and motor

6.1.7   Equipment Descriptions—Structural Steel and Boiler Casing

Structural Steel

Erect structural steel for:
    Stoker support steel
    Boiler
    Economizer
    Mechanical collector
    Air heater
    Scrubber
    Stack

Boiler Casing

Erect casing—generating bank, boiler, and penthouse

Waste Heat Boiler

6.2 General Scope of Field Work Required for Each 400,000 lb/h Waste Heat Boiler 400,000 lb/h Membrane Wall Boiler With Superheater Scope of Work-Field Erection

Pressure parts
Install steam and mud drums
  Steam drum trim piping
  Drum internals—remove and reinstall
Generating bank tubes
  Install generating tubes
  2-1/2″ OD×0.203″ wall thickness, swage, and roll
Ground assembly—headers, furnace, and water wall panels
  Assemble headers and wall panels, fit and weld water wall panels to headers
Erect headers and wall panels, fit and weld water wall panels
  Erect side wall, front, and rear panels-water wall; 2-31/32″ OD×0.203″ WT, BW (TIG)
  Fit and weld filler bar at tube welds
  Fit and weld filler joining membrane panels
Fit and weld water wall tubes; 2-31/32″ OD×0.203″ WT, BW (TIG)
Erect, fit and weld primary/secondary headers and superheater elements
  Primary superheater headers/coils
  10.75″ diameter inlet header
  20″ outlet header
  Weld superheater Tubes; 1.772″ OD×0.150″ WT, BW (TIG)
Burner system
  Burner and wind box
  Fan with drive
  Silencer
Soot Blowers
  Install rotary and retractable soot blowers and support bearings
Erect and install downcomers and steam code piping

6.3   400,000 lb/h Waste Heat Boiler Estimate Data

6.3.1   Drums, Generating Tubes, Headers, Side, Front, and Rear Wall Panels Sheet 1

Description	MH	Unit
Drums—Includes Straps and/or U-Bolts		
100,001–200,000 lbs	4.60	ton
200,001–300,000 lbs	3.80	ton
Steam drum trim piping	260.00	Lot
Drum internals: remove and reinstall	136.00	Boiler
Generating Bank Tubes		
Install generating tubes	0.50	EA
2′1/2″ Ends: expand tubes in steam and mud drums	0.44	End
Headers, Furnace, and Water Wall Panels		
Headers—loose	4.00	ton
Shop assembled wall panels with or without headers attached	2.50	ton
Field tube welding—over 2-1/2″ and including 3″ TIG (1500 PSI)	4.20	EA
Primary superheater headers/coils	4.20	ton
Fit and weld filler bar at tube welds	1.00	Space
Fit and weld filler joining membrane panels	0.60	LF
Field tube welding—over 1-1/2″ and including 2″ TIG (1500 PSI)	3.70	EA

6.3.2   Erect Headers and Panels, Weld Water Wall Tubes, Burner, Sootblower, Superheater Sheet 2

Description	MH	Unit
Erect Headers and Wall Panels, Fit and Weld Water Wall Panels		
Erect right, left front, rear, and roof panels	2.50	ton
Fit and weld filler bar at tube welds	1.00	Space
Fit and weld filler joining membrane panels	0.60	LF
Scaffolding and rigging	136.00	Boiler
Fit and weld water wall tubes; 2-31/32″ OD×0.203″ WT, BW (TIG)	4.20	EA
Erect, Fit, and Weld Primary/Secondary Headers and Superheated Elements		
Primary superheater headers/coils (27.5′ L×10′ W×10.5′ H) 48,281 lb per unit 4 EA	4.20	ton
10.75″ Diameter inlet header	4.00	ton
20″ Outlet header	4.00	ton
Weld superheater tubes; 1.772″ OD×0.150″ WT, BW (TIG)	3.70	EA
Burner System		
Burner and wind box	59.40	ton
Fan with drive	45.00	ton
Silencer	60.00	EA
Soot blower (24′ L×1′-6″ diameter×2′-0″ H 2570 lb per unit	64.00	EA

6.3.3 Downcomers and Code Piping Sheet 3

Description	MH	Unit
Erect and Install Down Comers and Code Piping		
Handle down comer; 12″ OD×0.562″ WT	1.32	LF
12″×0.562 WT CS BW	12.6	EA
Handle 12″ 300# BW valve	5.4	EA
Handle 10″ 300# BW valve	4.50	EA
Handle 8″ 300# BW spray control valves	7.20	EA
Handle 3″ 300# safety/control valves	2.70	EA
Handle 2″ 300# control valves—boiler	1.00	EA
10″×0.562 WT CS BW	10.5	EA
8″×0.562 WT CS BW	8.40	EA
3″×0.375 WT CS BW	1.50	EA
2″×0.375 WT CS BW	1.10	EA
4″ 300# Flanged air flow elements	3.60	EA
Handle 10″×0.562 pipe	1.10	LF
Handle 8″×0.562 pipe	0.88	LF
4″ 300# Bolt connection	1.60	EA
Handle 3″×0.375″ pipe	0.21	LF
Handle 2″×0.375″ pipe	0.18	LF

6.4 Waste Heat Boiler Installation Estimate

6.4.1 Drums, Generating Tubes, Headers, Side, Front, and Rear Panels Sheet 1

		Historical			Estimate		
Description	MH	Quantity	Unit		Quantity	Unit	BM
Install Steam and Mud Drums							**0**
Steam drum (6'-0" diameter × 48')	3.8	111	ton		0	ton	0
Mud drum (4'-6" diameter × 47')	4.6	63	ton		0	ton	0
Steam drum trim piping	260.0	1	Lot		0	Lot	0
Drum internals: remove and reinstall	136.0	1	Boiler		0	Boiler	0
Generating Bank Tubes 2-1/2" OD × 0.203" wall thickness, swage, and roll							**0**
Install generating tubes	0.50	2629	EA		0	EA	0
2'1/2" Ends: expand tubes in steam and mud drums	0.44	5258	End		0	End	0
Ground Assembly— Headers, Furnace, and Wall Panels Fit and weld water wall panels to headers							**0**
Lower side wall header—lower side wall panels							
Place lower side wall header	4.0	11	ton		0	ton	0
Place lower side wall panels-water wall	2.5	47	ton		0	ton	0
Field tube welding—over 2-1/2" TIG	4.2	156	EA		0	EA	0
Upper side wall header—upper side wall panels							
Place upper side wall header	4.0	9	ton		0	ton	0
Place upper side wall panels-water wall	2.5	47	ton		0	ton	0
Field tube welding—over 2-1/2" TIG	4.2	156	EA		0	EA	0
Place front wall header	4.0	5	ton		0	ton	0
Place front wall panels	2.5	24	ton		0	ton	0
Field tube welding—over 2-1/2" TIG	4.2	108	EA		0	EA	0
Place rear wall header	4.0	4	ton		0	ton	0
Place rear wall panels-water wall	2.5	24	ton		0	ton	0
Field tube welding—over 2-1/2" TIG	4.2	108	EA		0	EA	0

6.4.2   Erect Headers and Panels, Weld Water Wall Tubes, Burner, Sootblower, Superheater Sheet 2

Description	MH	Historical Quantity	Unit	Estimate Quantity	Unit	BM
Erect Headers and Wall Panels, Fit and Weld Wall Panels						**0**
Scaffolding and rigging	136.00	1.0	Boiler	0	Boiler	0
Erect right side wall panels-water wall	2.50	120.0	ton	0	ton	0
Fit and weld filler bar at tube welds	1.00	468.0	Space	0	Space	0
Fit and weld filler joining membrane panels	0.60	160.0	LF	0	LF	0
Erect left side wall panels-water wall	2.50	120.0	ton	0	ton	0
Fit and weld filler bar at tube welds	1.00	468.0	Space	0	Space	0
Fit and weld filler joining membrane panels	0.60	160.0	LF	0	LF	0
Erect front wall panels-water wall	2.50	120.0	ton	0	ton	0
Fit and weld filler bar at tube welds	1.00	432.0	Space	0	Space	0
Fit and weld filler joining membrane panels	0.60	156.0	LF	0	LF	0
Erect rear wall panels-water wall	2.50	120.0	ton	0	ton	0
Fit and weld filler bar at tube welds	1.00	432.0	Space	0	Space	0
Fit and weld filler joining membrane panels	0.60	156.0	LF	0	LF	0
Erect roof panels-water wall	2.50	90.0	ton	0	ton	0
Fit and weld filler bar at tube welds	1.00	588.0	Space	0	Space	0
Fit and weld filler joining membrane panels	0.60	173.0	LF	0	LF	0
Fit and weld water wall tubes	4.20	1645.8	EA	0	EA	**0**
Erect, Fit, and Weld Primary/ Secondary Headers						**0**
Primary superheater headers/coils	4.20	96.6	ton	0	ton	0
10.75″ Diameter inlet header	4.00	2.1	ton	0	ton	0
20″ Outlet header	4.00	5.0	ton	0	ton	0
Weld superheater tubes; 1.772″ OD	3.70	292.0	EA	0	EA	0
Burner System						**0**
Burner and wind box	59.40	4.7	ton	0	ton	0
Fan with drive	45.00	1.8	ton	0	ton	0
Silencer	60.00	1.0	EA	0	EA	0
Soot Blowers						**0**
Install rotary and retractable soot blowers						
Superheater—soot blower with pipe, valve, fitting	64.00	12.0	EA	0	EA	0
Economizer—soot blower with pipe, valve, fitting	64.00	4.0	EA	0	EA	0
Generating bank—soot blower with PVF	64.00	4.0	EA	0	EA	0

6.4.3 Downcomers and Code Piping Sheet 3

Description	MH	Historical Quantity	Unit	Estimate Quantity	Unit	BM
Erect and Install Down Comers and Code Piping						**0**
Handle down comers; 12″ OD×0.562 WT×40′	1.32	280.0	LF	0	LF	0
Down comers; 12″ OD×0.562 WT BW	12.60	14.0	EA	0	EA	0
Handle 12″ 300# BW main stop valve and Ck valve	5.4	2.0	EA	0	EA	0
12″×0.562 WT CS BW	12.60	4.0	EA	0	EA	0
Handle 10″ 300# BW feed water stop and Ck valve	4.50	2.0	EA	0	EA	0
10″×0.562 WT CS BW	10.50	4.0	EA	0	EA	0
Handle 8″ 300# BW spray control valves	7.20	4.0	EA	0	EA	0
8″×0.562 WT CS BW	8.40	8.0	EA	0	EA	0
Handle 3″ 300# safety valves	2.70	3.0	EA	0	EA	0
3″×0.375 WT CS BW	1.50	6.0	EA	0	EA	0
Handle 3″ 300# control valves—boiler	2.70	2.0	EA	0	EA	0
3″×0.375 WT CS BW	1.50	4.0	EA	0	EA	0
Handle 2″ 300# control valves—boiler	1.00	2.0	EA	0	EA	0
2″×0.375 WT CS BW	1.10	4.0	EA	0	EA	0
Handle 4″ 300# flanged flow elements	3.60	6.0	EA	0	EA	0
4″ 300# bolt up	1.60	12.0	EA	0	EA	0
Handle 10″×0.562 pipe	1.10	160.0	LF	0	LF	0
Handle 8″×0.562 pipe	0.88	160.0	LF	0	LF	0
Handle 4″×0.375 pipe	0.28	225.0	LF	0	LF	0
Handle 3″×0.375 pipe	0.21	160.0	LF	0	LF	0
Handle 2″×0.375 pipe	0.18	160.0	LF	0	LF	0

6.5   Waste Heat Boiler-Equipment Installation Man Hours

	Actual	Estimated
Facility—Biomass Plant Major Equipment	MH	BM
Install steam and mud drums	1103	0
Generating bank tubes	3628	0
Ground assembly—headers, furnace, and water wall panels	2692	0
Erect headers and wall panels, fit and weld water wall panels	4432	0
Fit and weld water wall tubes; 2-31/32″ OD×0.203″ WT, BW (TIG)	6912	0
Erect, fit, and weld primary/secondary headers and superheater elements	1514	0
Burner system	280	0
Sootblowers	1280	0
Erect and install downcomers and code piping	1272	0
Waste heat boiler-equipment and piping installation man hours	23114	0

6.6   General Scope of Field Work Required for Each Stoker

Vibrating hydro grate arranged for front ash discharge
Vibrating grate modules
Grate drive arrangement
Interior stoker seals
Seals for settings
Distribution spout air system
Fuel distributor mounting plate
Extension front arrangement
Stoker doors
Fuel distributors
Rotating air dampers
Balanced dampers
Overfire air turbulence system

6.7   Stoker Estimate Data

6.7.1   Vibrating Grate Modules, Flues and Ducts, Fuel Distributors Sheet 1

Description	MH	Unit
Stoker-Detroit reciprocating grate	22.00	ton
Flues and ducts, including supports, hoppers, doors, expansion joints, and dampers	40.00	ton
Spray nozzle assembly and reducer	72.00	ton
Air heater and nozzle	30.00	ton
Fuel distributors	24.00	ton

6.8 Stoker Equipment Installation Estimate

6.8.1 Modules, Flues and Ducts, Fuel Distributors Sheet 1

Description	Historical			Estimate		
	MH	Quantity	Unit	Quantity	Unit	BM
Stoker						**0**
Vibrating grate modules	22	37	ton	0	ton	0
Four shop-assembled grate modules complete with support steel to grate. Each module will be complete with the following components: One set shop-installed grate Four H-Type "K" thermocouple assemblies with clamps, fitting, IC wire to junction box. One stationary frame for support of the vibrating frame One set of structural legs One set flexing straps for support of the vibrating grid section						
Grate Drive Arrangement						
One drive arrangement per stoker module consisting of one 7.5 HP motor, belt drive, drive guard, pillow block bearings, and drive shaft with eccentric linked to vibrating frame						
Interior Stoker Seals						
One set "T" seals between stationary and vibrating sections of the stoker						
Seals for Settings						
One set high-pressure air seals at the sides, front, and rear of the setting. Seals attach to the stationary frames and the boiler side and rear wall headers install and vulcanize expansion joint						
Distribution Spout Air System						
One set high-pressure air ducting with bracing, including manual dampers to serve refuse distributor spouts	40	4	ton	0	ton	0
Fuel Distributor Mounting Plate						
Reinforced steel mounting plates to support the fuel distributors						

6.8.2 Fuel Distributors, Dampers, Overfire Air Turbulence System Sheet 2

Description	Historical			Estimate		
	MH	Quantity	Unit	Quantity	Unit	BM
Extension Front Arrangement						
Extension front is designed with removable steel front panels						
Front panels, side panels, and roof panels						
Stoker Doors						
Access panels in sides of stationary frame						
Refractory-lined, hinged access doors grate module mounted in extension front panels						
Fuel Distributors						
Six spouts located	24	19	EA	0	EA	0
Rotating Air Dampers						
Six rotating air damper assemblies	40	6	EA	0	EA	0
Dampers are driven by one 1/3 H gear motor						
Balanced Dampers						
Six balanced damper assemblies	40	6	EA	0	EA	0
Overfire Air Turbulence System						
One Detroit Stoker Company overfire turbulence system eight high-pressure overfire air heaters and nozzles	30	3	EA	0	EA	0
One manual control damper	40	6	EA	0	EA	0

6.9 Stoker-Equipment Installation Man Hours

	Actual	Estimated
Facility—Biomass Plant Major Equipment	**MH**	**BM**
Stoker	2236	0
Stoker-equipment and piping installation man hours	2236	0

Economizer

6.10   General Scope of Field Work Required for Each Economizer

Install support steel and platforms
   Set one modular economizer complete with access platforms
   Install inlet duct
   Install outlet duct

6.11   Economizer Estimate Data

6.11.1   Support Steel, Modular Sheet 1

Description	MH	Unit
Economize Support Steel		
Main steel	24.00	ton
Shim/set base plates	2.00	EA
Make-up base plates	2.60	EA
Grout base plates	4.00	EA
Bolted connection	1.50	EA
Platform framing	0.15	SF
Grating	0.20	SF
Handrail	0.25	LF
Economizer shop assembled	2.20	ton
Flues and ducts, including supports, hoppers, doors, expansion joints, and dampers	40.00	ton
Field joint	0.50	LF

6.12   Economizer Equipment Installation Estimate

6.12.1   Support Steel, Modular Sheet 1

Description	Historical			Estimate		
	MH	Quantity	Unit	Quantity	Unit	BM
Economize Support Steel						**0**
W 8×31×23′×4EA	24.00	1.4	ton	0	ton	0
Shim/set base plates	2.00	4.0	EA	0	EA	0
Make-up base plates	2.60	4.0	EA	0	EA	0
Grout base plates	4.00	4.0	EA	0	EA	0
W 8×31×20′×4EA horizontal beam	24.00	1.2	ton	0	ton	0
W 8×31×20′×4EA vertical	24.00	1.2	ton	0	ton	0
Bolted connection	1.50	8.0	EA	0	EA	0
Platform framing	0.15	400.0	SF	0	SF	0
Grating	0.20	400.0	SF	0	SF	0
Handrail	0.25	80.0	LF	0	LF	0
Economizer modular (38′ L×10′ W×10′-6″ H)	2.20	58.0	ton	0	ton	**0**
Inlet Duct						**0**
Duct elbow	40.00	1.1	ton	0	ton	0
Duct transition	40.00	2.8	ton	0	ton	0
Field joint	0.50	120.0	LF	0	LF	0
Outlet Duct						**0**
Expansion joint	40.00	1.0	ton	0	ton	0
Duct transition	40.00	2.0	ton	0	ton	0
Field joint	0.50	120.0	LF	0	LF	0

6.13   Economizer-Equipment Installation Man Hours

Facility—Biomass Plant Major Equipment	Actual	Estimated
	MH	BM
Economize support steel	300	0
Economizer modular (38′ L×10′ W×10′-6″ H)	128	0
Inlet duct	216	0
Outlet duct	180	0
Economizer-equipment installation man hours	**824**	**0**

6.14 General Scope of Field Work Required for Each Precipitator

Precipitator Supports: All structural steel slide plates.

One PPC Model 59R-1534-5712P modular electrostatic precipitator including all collecting plates, rigid discharge electrodes, roof sections, insulator compartments, access doors, and all internal components

Install hoppers

Collecting plates—both top and bottom alignment guides, stiffeners, and mountings will maintain alignment of the collecting plates

Electromagnetic uplift-gravity impact rappers

Discharge electrode frames

Step-up transformers/rectifiers

High tension support insulators

Nozzles; flanged inlet and outlet nozzles

Roof equipment enclosure-mounted on roof

Purge blower control panel

6.15 Precipitator Estimate Data

6.15.1 Module, Collecting Plates, Rappers Rectifiers, Electrodes, Panel Sheet 1

Description	MH	Unit
One PPC Model 59R-1534-5712P modular electrostatic precipitator	5.00	ton
Install collecting plates	3.50	EA
Install rappers	3.00	EA
Set transformer rectifier sets	14.00	EA
Discharge electrode frames	34.00	EA
Roof equipment enclosure-mounted on roof	60.00	EA
Purge blower control panel	40.00	EA

6.16   Equipment Installation Estimate

6.16.1   Module, Collecting Plates, Rappers Rectifiers, Electrodes, Panel Sheet 1

	Historical			Estimate		
	MH	Quantity	Unit	Quantity	Unit	BM
One PPC Model 59R-1534-5712P modular ESP	5.00	174.4	ton	0	ton	0
Install collecting plates	3.5	228.6	EA	0	EA	0
Install rappers	3.00	109.8	EA	0	EA	0
Set transformer rectifier sets	14.00	5.0	EA	0	EA	0
Discharge electrode frames	34.00	10.0	EA	0	EA	0
Roof equipment enclosure-mounted on roof	60.00	1.0	EA	0	EA	0
Purge blower control panel	40.00	1.0	EA	0	EA	0

6.17   Precipitator-Equipment Installation Man Hours

	Actual	Estimated
Facility—Biomass Plant Major Equipment	MH	BM
One PPC Model 59R-1534-5712P modular ESP	872	0
Install collecting plates	800	0
Install rappers	330	0
Set transformer rectifier sets	70	0
Discharge electrode frames	340	0
Roof equipment enclosure-mounted on roof	60	0
Purge blower control panel	40	0
Precipitator-equipment installation man hours	**2512**	**0**

6.18   General Scope of Field Work Required for Each Mechanical Collector

Install collector with support steel and access platforms
One Warren Environmental Model 288-12-A unit
Set Hopper Cones

6.19   Mechanical Collector Estimate Data

6.19.1   Support Steel, Module and Hopper Cones Sheet 1

Description	MH	Unit
Mechanical Collector Support Steel		
Main steel	24.00	ton
Bolted connection	1.50	EA
Platform framing	0.15	SF
Grating	0.20	SF
Handrail	0.25	LF
One Warren environmental model	40.00	ton
288-12-A unit		
Set hopper cones	20.00	ton

6.20   Equipment Installation Estimate

6.20.1   Support Steel, Module and Hopper Cones Sheet 1

		Historical			Estimate		
	MH	**Quantity**	**Unit**	**Quantity**	**Unit**	**BM**
Mechanical Collector Support Steel						**0**
W 8×31×23′×4EA	24.00	1.0	ton	0	ton	0
W 8×31×20′×4EA horizontal beam	24.00	0.8	ton	0	ton	0
W 8×31×20′×4EA vertical	24.00	0.8	ton	0	ton	0
Bolted connection	1.50	8.0	EA	0	EA	0
Platform framing	0.15	48.0	SF	0	SF	0
Grating	0.20	48.0	SF	0	SF	0
Handrail	0.25	28.0	LF	0	LF	0
One Warren environmental model 288-12-A unit	40.00	5.0	ton	0	ton	0
Set Hopper Cones	20.00	34.4	EA	0	EA	0

6.21   Mechanical Collector-Equipment Installation Man Hours

	Actual	Estimated
Facility—Biomass Plant Major Equipment	**MH**	**BM**
Mechanical collector support steel	96	0
One Warren environmental model 288-12-A unit	200	0
Set hopper cones	688	0
Mechanical collector-equipment installation man hours	**984**	**0**

6.22   General Scope of Field Work Required for Each CO and No$_x$ Removal System

One Model C1 × 192(H) CO
One Model N2 × 256 (H) No$_x$ removal catalyst system
CO catalyst removal system will have the following design features:
  Catalyst layers—1
  CO catalyst layer depth (inches)—2.0
NO$_x$ catalyst removal system will have the following design features:
  No$_x$ reduction reagent—Anhydrous ammonia
  Flow distribution devices—as required
  Catalyst layers—2
  Ammonia injection grid—1
  Ammonia injection grid with adjustable lances and distribution headers with nozzles
  Catalyst will be mounted in honeycomb blocks—18″ × 22.5″
  Catalyst housing shell will be seal welded to form a gas tight structure
  Shell—25′-9″ L × 25′-0″ W × 45′-0″ H

6.23   CO and No$_x$ Removal Estimate Data

6.23.1   CO and No$_x$ Removal System Sheet 1

Description	MH	Unit
One model C1 × 192(H) CO	20.00	ton
Assemble shell	20.00	ton
Field joint	0.40	LF
Ammonia Grid		
Ammonia grid module	20.00	ton
Field joint	0.40	LF
Remove shipping angles	40.00	Module
Install tie plate	40.00	Module
SCR Seal Frame		
Seal frame	20.00	ton
Catalyst Modules		
Catalyst modules	10.00	EA
Install catalyst	4.00	EA

6.24 Equipment Installation Estimate

6.24.1 CO and No$_x$ Removal System Sheet 1

Facility Class—Biomass Plant Major Equipment	Historical			Estimate		
	MH	Quantity	Unit	Quantity	Unit	BM
One model C1 × 192(H) CO	20.00	6.0	ton	0	ton	**0**
One model N2 × 256 (H) No$_x$ **removal catalyst system**						**0**
Catalyst housing-seal weld - field joint	0.40	486.0	LF	0	LF	0
Assemble shell	20.00	24.5	ton	0	ton	0
Ammonia Grid						**0**
Ammonia grid module	20.00	6.0	ton	0	ton	0
Field joint	0.40	204.0	LF	0	LF	0
Remove shipping angles	40.00	1.0	Module	0	Module	0
Install tie plate	40.00	1.0	Module	0	Module	0
SCR Seal Frame						**0**
Seal frame	20.00	6.0	ton	0	ton	0
Catalyst Modules						**0**
Catalyst modules	10.00	8.0	EA	0	EA	0
Install catalyst	4.00	80.0	EA	0	EA	0

6.25 CO and No$_x$ Removal System-Equipment Installation Man Hours

Facility—Biomass Plant Major Equipment	Actual	Estimated
	MH	BM
One model C1 × 192(H) CO	120	0
One model N2 × 256 (H) No$_x$ removal catalyst system	684	0
Ammonia grid	282	0
SCR seal frame	120	0
Catalyst modules	400	0
CO and No$_x$ removal system-equipment installation man hours	**1606**	**0**

6.26   General Scope of Field Work Required for Each Wet Scrubber

Wet Scrubber (22'-0" Diameter × 39')

Knocked Down:

Unit Shipped broken down in half-moon sections—12' W × 24' L × 12' H.

All sections are flanged for field bolting.

Seal seams on inside only; apply 2" wide fiberglass matte strips followed by a 3" wide Nexus veil strip coated with fiberglass gel.

Six spray headers mounted on nozzles.

There is a 6'-0" tall bed of polypropylene random dumped packing.

Field install mist eliminator modules held down by FRP Beams.

6.27   Wet Scrubber Estimate Data

6.27.1   Wet Scrubber Sheet 1

Description	MH	Unit
Wet scrubber (22'-0" diameter × 39')	30.00	ton
Temporary field bolting and seal seam	0.65	LF
264' Diameter mist eliminator module	176.00	Module
264' Diameter vessel packing—hold down plate	72.00	EA

6.28   Equipment Installation Estimate

6.28.1   Wet Scrubber Sheet 1

	Historical			Estimate		
	MH	Quantity	Unit	Quantity	Unit	BM
Wet scrubber (22'-0" Diameter × 39')	30.00	10.0	ton	0	ton	0
Temporary field bolting and seal seam	0.65	216.0	LF	0	LF	0
264" Diameter mist eliminator module	176.00	1.0	Module	0	Module	0
264" Diameter vessel packing—hold down plate	72.00	1.0	EA	0	EA	0

6.29   Wet Scrubber-Equipment Installation Man Hours

	Actual	Estimated
	MH	BM
Wet scrubber (22′-0″ Diameter × 39′)	300	0
Temporary field bolting and seal seam	140	0
264″ Diameter mist eliminator module	176	0
264″ Diameter vessel packing—hold down plate	72	0
Wet scrubber-equipment installation man hours	**688**	**0**

6.30   General Scope of Field Work Required for Each Tubular Air Heater Modules

Erect support steel
Set Tubular air heater modules
Set plenums
Erect interconnecting ductwork

6.31   Tubular Air Heater Estimate Data

6.31.1   Tubular Air Heater Modules Sheet 1

Description	MH	Unit
Support Steel		
W 8×31×30′×3 EA	24.00	ton
Structural; legs with base plates	4.00	EA
Shim/set base plates	2.00	EA
Make-up base plates	2.60	EA
Grout base plates	4.00	EA
W 8×31×42′×3 EA horizontal beam	24.00	ton
Bolted connection	1.50	EA
Tubular air heater modules	2.70	ton
Plenum	11.00	ton
Erect Interconnecting Ductwork		
Duct transition, reducer, elbow, and sections	32.00	ton
Bolted connection	0.26	EA

6.32   Equipment Installation Estimate

6.32.1   Tubular Air Heater Modules Sheet 1

	Historical			Estimate		
	MH	Quantity	Unit	Quantity	Unit	BM
Support Steel						**0**
W 8×31×30′×3 EA	24.00	1.4	ton	0	ton	0
Structural; legs with base plates	4.00	4.0	EA	0	EA	0
Shim/set base plates	2.00	4.0	EA	0	EA	0
Make-up base plates	2.60	4.0	EA	0	EA	0
Grout base plates	4.00	4.0	EA	0	EA	0
W 8×31×42′×3 EA horizontal beam	24.00	2.0	ton	0	Ton	0
Bolted connection	1.50	19.5	EA	0	EA	0
Tubular air heater modules (23′×11′-7″×9′-0″)	2.70	146.0	ton	0	ton	**0**
Plenum (10′ diameter×43′) 2 EA 6000 lb per unit	11.00	14.5	ton	0	ton	**0**
Erect Interconnecting Ductwork						**0**
Duct transition	32.00	2.0	ton	0	ton	0
Bolted connection-duct transi- tion to air heater module	0.26	92.0	EA	0	EA	0
Duct sections	32.00	4.0	ton	0	ton	0
Duct reducer	32.00	1.0	ton	0	ton	0
Duct elbow	32.00	3.0	ton	0	ton	0
Bolted connection-duct sections	0.26	440.0	EA	0	EA	0

6.33   Tubular Air Heater Modules-Equipment Installation Man Hours

Facility—Biomass Plant Major	Actual	Estimated
Equipment	MH	BM
Support steel	160	0
Tubular air heater modules	394	0
Plenum	160	0
Erect interconnecting ductwork	458	0
Tubular air heater modules-equipment installation man hours	**1172**	**0**

6.34   General Scope of Field Work Required for Each Stack

One filament wound FRP stack 10′ I.D. × 128′ H with open bottom with flange, open top with 10′×9′-4″ reducer. Stack has following features:

Four FRP body flanges

Three carbon steel galvanized guy/brace points (4) clips each at 24′ elevation from bottom of stack

Six carbon steel galvanized lifting lugs

Sixty FRP lightening protection clips

6.35   Stack Estimate Data

6.35.1   Stack Sheet 1

Description	MH	Unit
Stack	20.00	ton
Bolted connection	0.26	EA

6.36   Equipment Installation Estimate

6.36.1   Stack Sheet 1

	Historical			Estimate		
	MH	Quantity	Unit	Quantity	Unit	BM
Stack 10′ diameter×43′ 6000 lb per unit 3 EA	20.00	9.0	ton	0	ton	0
Bolted connection-sections	0.32	128.0	EA	0	EA	0

6.37   Stack-Equipment Installation Man Hours

	Actual	Estimated
Facility—Biomass Plant Major Equipment	MH	BM
Stack	180	0
Bolted connection-sections	40	0
Stack-equipment installation man hours	**220**	**0**

6.38   General Scope of Field Work Required for Fuel Feeding Equipment

Metering bins
Install six metering bin screws. Installation includes support steel. Metering bins will set on load cells. Install 24 load cells
Boiler drag chain
Install one boiler drag chain above the metering bins
Install 8'-0" × 6'-0" enclosure and six rack and pinions, slide gates
Solid fuel is fed to the boiler in six feed points, all on the front wall
Fuel chutes
Install metering bin discharge chutes, expansion joints to stoker windswept spouts, support bins, and drag chains for all six bins

6.39   Fuel Feeding Equipment Estimate Data

6.39.1   Fuel Feeding Equipment Sheet 1

Description	MH	Unit
Metering Bins		
Support steel	24.00	ton
Install metering bin screws	19.40	EA
Metering bins on load cells	130.00	EA
Boiler Drag Chain		
Boiler drag chain section	120.00	Section
Slide gates	44.40	ton
Steel enclosure	20.00	EA
Rack and pinions, including chain wheel actuators between the meters	16.00	EA
Fuel Chutes		
Metering bin discharge chutes	40.00	ton
Expansion joints to stoker	40.00	ton
Support bins and drag chains	60.00	EA

6.40 Equipment Installation Estimate

6.40.1 Fuel Feeding Equipment Sheet 1

	Historical			Estimate		
	MH	Quantity	Unit	Quantity	Unit	BM
Metering Bins						**0**
Metering bins support steel	24.00	6.8	ton	0	ton	0
Install metering bin screws	19.40	6.0	EA	0	EA	0
Set metering bins on load cells (17′ L × 4′ W × 10′ H)	130.00	6.0	EA	0	EA	0
Boiler Drag Chain						**0**
Boiler drag chain (62′ L × 5′6″ W × 5′-0″ H) and one each head and tail section	120.00	4.0	ton	0	ton	0
Slide gates (2′ L × 5′ W × 1′ H)	44.40	6.0	ton	0	ton	0
8′ × 6′ Steel enclosure	20.00	1.0	EA	0	EA	0
Six rack and pinions	16.00	6.0	EA	0	EA	0
Fuel Chutes						**0**
Install metering bin discharge chutes-from bins to stoker	11.72	18.8	ton	0	ton	0
Expansion joints to stoker	40.00	6.0	ton	0	ton	0
Support bins and drag chains	60.00	6.0	EA	0	EA	0

6.41 Fuel Feeding Equipment-Equipment Installation Man Hours

Facility—Biomass Plant Major Equipment	Actual	Estimated
	MH	BM
Metering bins	1060	0
Boiler drag chain	862	0
Fuel chutes	820	0
Fuel feeding equipment-equipment installation man hours	**2742**	**0**

6.42   General Scope of Field Work Required for Bottom Ash Drag System

Equipment consists of refractory-lined bottom ash chute and one submerged ash drag chain conveyor.

Bottom ash chute is 3/8″ carbon steel, 34′-0″ L × 3′-0″ W × 1′-0″ H.

Submerged ash drag conveyor is 3/8″ carbon steel × 66′-0″ L with outlet chute.

Shipped in sections:
    Head section—motor and drive shipped attached to head
    Neck section
    Transition section
    Straight section
    Tail section-take-up shipped assembled

Scope of work:
    Assembly of housing
    Assembly of drag chain
    Installation of drive units
    Pneumatic/hydraulic take-up units, including air piping
    Discharge chutes
    Grouting
    Refractory-lined chute on inlet of submerged ash drag
    All water piping to and from submerged ash drag and drains

6.43   Bottom Ash Drag System Estimate Data

6.43.1   Bottom Ash Drag System Sheet 1

Description	MH	Unit
Bottom Ash Drag System		
Drive, head and neck section	120.00	Section
Refractor lined ash chute	21.30	ton
Grout	280.00	Lot
Water piping	200.00	LF

6.44   Equipment Installation Estimate

6.44.1   Bottom Ash Drag System Sheet 1

	Historical			Estimate		
	MH	Quantity	Unit	Quantity	Unit	BM
Bottom Ash Drag System						**0**
Drive, head and neck section; lot 7000 lb	120.00	3.0	Section	0	Section	0
Transition and trough section; lot 12,000 lb	120.00	2.0	Section	0	Section	0
Trail and trough section; lot 12,000 lb	120.00	2.0	Section	0	Section	0
Refractor lined ash chute; lot 15,000 lb	21.30	7.5	ton	0	ton	0
Grout	120.00	1.0	Lot	0	Lot	0
Water piping	120.00	1.2	LF	0	LF	0

6.45   Bottom Ash Drag System-Equipment Installation Man Hours

	Actual	Estimated
Facility—Biomass Plant Major Equipment	MH	BM
Bottom ash drag system	1264	0
Bottom ash drag system-equipment installation man hours	**1264**	**0**

6.46   General Scope of Field Work Required for ID, OFA, FD, and Spout Air Fan

Fans

Perform all millwright work for the four fans; alignment and grouting.

ID Fan, OFA Fan, and FD Fan

Install one Process Barron 2000 HP ID Fan

Install one Process Barron 800 HP OFA Fan

Install one Process Barron 400 HP FD Fan

Work scope:

Set housing and inlet box

Set shaft and inlet cones

Install motor sole plates

Set inlet damper and damper drive

Set motor

Set inlet silencer

Lubrication

Laser alignment

Final dowelling

Grouting of bearings and motor

Spout air fan

Install one industrial air products 200 HP Spout Air Fan

Set unitary base housing, motor, and inlet box

Set shaft and inlet cones

Set two ambient air inlet dampers and damper drive

Set FGR inlet damper and damper drive

Set inlet silencer

Lubrication

Laser alignment

Final dowelling

Grouting of bearings and motor

6.47 ID, OFA, FD, and Spout Air Fan Estimate Data

6.47.1 ID, OFA, FD, and Spout Air Fan Sheet 1

Description	MH	Unit
Fans		
Install sole and base plates	2.0	EA
Attach fan to foundation-bolted connection	0.42	EA
ID Fan		
Set fan housing-split	12.0	EA
Set impeller assembly/shaft	16.0	EA
Alignment fan bearings	40.0	EA
Seal weld housing inside/outside	48.0	JOB
Weld inlet cones and stud rings	12.0	JOB
Weld shaft seal stud rings	12.0	JOB
Coupling alignment	32.0	JOB
Install damper	40.0	EA
OFA Fan		
Set fan housing-split	6	EA
Set impeller assembly/shaft	8	EA
Alignment fan bearings	16	EA
Seal weld housing inside/outside	20	JOB
Weld inlet cones and stud rings	6	JOB
Weld shaft seal stud rings	6	JOB
Coupling alignment	14	JOB
Install damper	16	EA
FD Fan		
Set fan housing-split	4	EA
Set impeller assembly/shaft	4	EA
Alignment fan bearings	10	EA
Seal weld housing inside/outside	12	JOB
Weld inlet cones and stud rings	4	JOB
Weld shaft seal stud rings	4	JOB
Coupling alignment	6	JOB
Install damper	10	EA

6.47.2 ID, OFA, FD, and Spout Air Fan Sheet 2

Description	MH	Unit
Set and Assemble Spout Air Fan		
Set fan housing-split	8.00	EA
Set impeller assembly/shaft	12.00	EA
Alignment fan bearings	20.00	EA
Seal weld housing inside/outside	24.00	JOB
Weld inlet cones and stud rings	8.00	JOB
Weld shaft seal stud rings	8.00	JOB
Coupling alignment	14.00	JOB
Install damper	18.00	EA
Steam coil air heater	28.50	ton

6.48 Equipment Installation Estimate

6.48.1 ID, OFA, FD, and Spout Air Fan Sheet 1

	Historical			Estimate			
	MH	Quantity	Unit	Quantity	Unit	BM	MW
Set and Assemble ID Fan						0	0
Install sole and base plates	2.0	6.0	EA	0	EA		0
Set fan housing-split	12.0	2.0	EA	0	EA	0	
Set impeller assembly/shaft	16.0	1.0	EA	0	EA	0	
Alignment fan bearings	40.0	1.0	EA	0	EA		0
Seal weld housing inside/ outside	48.0	1.0	JOB	0	JOB	0	
Weld inlet cones and stud rings	12.0	1.0	JOB	0	JOB	0	
Weld shaft seal stud rings	12.0	1.0	JOB	0	JOB	0	
Coupling alignment	32.0	1.0	JOB	0	JOB		0
Install damper	40.0	1.0	EA	0	EA	0	
Attach fan to foundation-bolted Connection	0.42	24.0	EA	0	EA	0	
Set and Assemble OFA Fan						0	0
Install sole and base plates	2.0	6.0	EA	0	EA		0
Set fan housing-split	6	2.0	EA	0	EA	0	
Set impeller assembly/shaft	8	1.0	EA	0	EA	0	
Alignment fan bearings	16	1.0	EA	0	EA		0
Seal weld housing inside/ outside	20	1.0	JOB	0	JOB	0	
Weld inlet cones and stud rings	6	1.0	JOB	0	JOB	0	
Weld shaft seal stud rings	6	1.0	JOB	0	JOB	0	
Coupling alignment	14	1.0	JOB	0	JOB		0
Install damper	16	1.0	EA	0	EA	0	
Attach fan to foundation-bolted connection	0.42	24.0	EA	0	EA	0	
Set and Assemble FD Fan						0	0
Install sole and base plates	2.0	6.0	EA	0	EA		0
Set fan housing-split	4	2.0	EA	0	EA	0	
Set impeller assembly/shaft	4	1.0	EA	0	EA	0	
Alignment fan bearings	10	1.0	EA	0	EA		0
Seal weld housing inside/ outside	12	1.0	JOB	0	JOB	0	
Weld inlet cones and stud rings	4	1.0	JOB	0	JOB	0	
Weld shaft seal stud rings	4	1.0	JOB	0	JOB	0	
Coupling alignment	6	1.0	JOB	0	JOB		0
Install damper	10	1.0	EA	0	EA	0	
Attach fan to foundation- bolted connection	0.42	24.0	EA	0	EA	0	

6.48.2 ID, OFA, FD, and Spout Air Fan Sheet 2

	Historical			Estimate			
	MH	Quantity	Unit	Quantity	Unit	BM	MW
Set and Assemble Spout Air Fan						**0**	**0**
Install sole and base plates	2.0	6.0	EA	0	EA		0
Set fan housing-split	8	2.0	EA	0	EA	0	
Set impeller assembly/shaft	12	1.0	EA	0	EA	0	
Alignment fan bearings	20	1.0	EA	0	EA		0
Seal weld housing inside/outside	24	1.0	JOB	0	JOB	0	
Weld inlet cones and stud rings	8	1.0	JOB	0	JOB	0	
Weld shaft seal stud rings	8	1.0	JOB	0	JOB	0	
Coupling alignment	14	1.0	JOB	0	JOB		0
Install damper	18	2.0	EA	0	EA	0	
Attach fan to foundation-Bolted Connection	0.42	24.0	EA	0	EA	0	
Steam Coil Air Heater	28.50	2.8	ton	0	ton	**0**	

6.49 ID, OFA, FD, and Spout Air Fan-Equipment Installation Man Hours

	Actual		Estimated	
Facility—Biomass Plant Major Equipment	MH	BM	MW	MH
Set and assemble ID fan	246	0	0	0
Set and assemble OFA fan	120	0	0	0
Set and assemble FD fan	80	0	0	0
Set and assemble spout air fan	160	0	0	0
Steam coil air heater	80	0		0
ID, OFA, FD, and Spout Air Fan-Equipment Installation Man Hours	686	0	0	0

Ductwork

6.50   General Scope of Field Work Required for Ductwork

Ductwork and breeching
   Install ductwork and breeching and expansion joints. Fit-up and seal welding from inside. Fit-up flanges tack welded to sections to install and bolt field joints.
   Ductwork will rest on sliding supports or hang from supports.

6.51   ID, OFA, FD, Underfire, and Overfire Ductwork Estimate Data

6.51.1   ID, OFA, FD, Underfire, and Overfire Ductwork Sheet 1

Description	MH	Unit
ID, OFA, and FD Ductwork		
Duct transition, reducer, elbow, and sections	32.00	ton
Bolted connection	0.26	EA
Field joint—weld flange	0.35	LF
Expansion joint	40.00	ton
Duct support	40.00	ton

6.51.2   Underfire and Overfire Ductwork Sheet 2

Description	MH	Unit
Underfire and Overfire Ductwork		
Duct transition, reducer, elbow, and sections	32.00	ton
Bolted connection	0.26	EA
Field joint—weld flange	0.35	LF
Expansion joint	40.00	ton
Duct support	40.00	ton

6.52 Equipment Installation Estimate

6.52.1 ID, OFA, FD, Underfire, and Overfire Ductwork Sheet 1

	Historical			Estimate		
	MH	Quantity	Unit	Quantity	Unit	BM
ID Ductwork						**0**
Transition duct section	32.00	1.5	ton	0	ton	0
Bolted connection-duct transition to ID fan	0.26	92.0	EA	0	EA	0
Duct sections	32.00	8.0	ton	0	ton	0
Duct elbow	32.00	2.0	ton	0	ton	0
Field joint—weld flange	0.35	320.0	LF	0	LF	0
Bolted connection-duct sections	0.26	640.0	EA	0	EA	0
Expansion joint	40.00	2.0	ton	0	ton	0
Duct support	40.00	3.0	ton	0	ton	0
OFA Ductwork						**0**
Transition duct section	32.00	1.0	ton	0	ton	0
Bolted connection-duct transition to ID fan	0.26	64.0	EA	0	EA	0
Duct sections	32.00	6.0	ton	0	ton	0
Duct elbow	32.00	2.0	ton	0	ton	0
Field joint—weld flange	0.35	280.0	LF	0	LF	0
Bolted connection-duct sections	0.26	512.0	EA	0	EA	0
Expansion joint	40.00	2.0	ton	0	ton	0
Duct support	40.00	2.0	ton	0	ton	0
FD Ductwork						**0**
Transition duct section	32.00	0.8	ton	0	ton	0
Bolted connection-duct transition to ID fan	0.26	56.0	EA	0	EA	0
Duct sections	32.00	6.0	ton	0	ton	0
Duct elbow	32.00	2.0	ton	0	ton	0
Field joint—weld flange	0.35	220.0	LF	0	LF	0
Bolted connection-duct sections	0.26	432.0	EA	0	EA	0
Expansion joint	40.00	2.0	ton	0	ton	0
Duct support	40.00	1.5	ton	0	ton	0

6.52.2   Underfire and Overfire Ductwork Sheet 2

	Historical			Estimate		
	MH	Quantity	Unit	Quantity	Unit	BM
Underfire Air Duct						**0**
Transition duct section	32.00	1.5	ton	0	Ton	0
Bolted connection—duct transition to ID fan	0.26	92.0	EA	0	EA	0
Duct sections	32.00	6.0	ton	0	ton	0
Duct elbow	32.00	2.0	ton	0	ton	0
Field joint—weld flange	0.35	280.0	LF	0	LF	0
Bolted connection—duct sections	0.26	560.0	EA	0	EA	0
Expansion joint	40.00	2.0	ton	0	ton	0
Duct support	40.00	3.0	ton	0	ton	0
Overfire Air Duct						**0**
Transition duct section	32.00	1.5	ton	0	ton	0
Bolted connection—duct transition to ID fan	0.26	92.0	EA	0	EA	0
Duct sections	32.00	8.0	ton	0	ton	0
Duct elbow	32.00	2.0	ton	0	ton	0
Field joint—weld flange	0.35	341.0	LF	0	LF	0
Bolted connection—duct sections	0.26	640.0	EA	0	EA	0
Expansion joint	40.00	2.0	ton	0	ton	0
Duct support	40.00	3.0	ton	0	ton	0

6.53   ID, OFA, FD, Underfire, and Overfire Ductwork-Equipment Installation Man Hours

	Actual	Estimated
Facility—Biomass Plant Major Equipment	**MH**	**BM**
ID ductwork	870	0
OFA ductwork	696	0
FD ductwork	624	0
Underfire air duct	772	0
Overfire air duct	879	0
ID, OFA, FD, underfire, and overfire ductwork-equipment installation man hours	3840	0

Structural Steel and Boiler Casing

6.54 General Scope of Field Work Required for Structural Steel and Boiler Casing

Erect boiler structural steel
    Column line "1" (west)
    Column line "6" (east)
    Column line "7" (east)
    Column line "A" (north)
    Column line "B" (north)
    Column line "F" (south)
    Column line "G" (south)
Install platform, grating and handrail at El. 100'-0" to 184'-0"
Install Stair No. 1 and no. 2
Erect penthouse steel
Erect casing—generating bank, boiler, and penthouse

6.55 Structural Steel and Boiler Casing Estimate Data

6.55.1 Structural Steel Sheet 1

Description	MH	Unit
Erect Boiler Structural Steel (Stairs, Platforms, Grating, and Handrails)		
Main steel	24.00	ton
Platform framing	0.15	SF
Grating	0.20	SF
Handrail	0.25	LF
Erect penthouse steel	24.00	ton
Install Platform, Grating, and Handrail at El. 100'-0" to 184'0"		
Platform framing	0.15	SF
Grating	0.20	SF
Handrail	0.25	LF
Install Stair No. 1 and No. 2		
Structural steel	24.00	ton
Platform framing	0.15	SF
Grating	0.20	SF
Handrail	0.25	LF
Stair treads	0.85	EA

6.55.2 Boiler Casing Sheet 2

Description	MH	Unit
Erect casing—generating bank, boiler, and penthouse	62	ton

6.56   Equipment Installation Estimate

6.56.1   Structural Steel Sheet 1

	Historical			Estimate		
	MH	Quantity	Unit	Quantity	Unit	BM
Erect Boiler Structural Steel						**0**
Column line "1" (west)	24.00	51.1	ton	0	ton	0
Column line "6" (east)	24.00	10.7	ton	0	ton	0
Column line "7" (east)	24.00	10.7	ton	0	ton	0
Column line "A" (north)	24.00	14.0	ton	0	ton	0
Column line "B" (north)	24.00	37.9	ton	0	ton	0
Column line "F" (south)	24.00	30.2	ton	0	ton	0
Column line "G" (south)	24.00	37.2	ton	0	ton	0
Erect penthouse steel	24.00	45.8	ton	0	ton	0
Install Platform, Grating, and Handrail at El. 100'-0" to 184'-0"						**0**
Platform framing	0.15	15488.0	SF	0	SF	0
Grating	0.20	15488.0	SF	0	SF	0
Handrail	0.25	2816.0	LF	0	LF	0
Install Stair No. 1 and No. 2						**0**
Structural steel	24.00	14.4	ton	0	ton	0
Platform framing	0.15	1440.0	SF	0	SF	0
Grating	0.20	1440.0	SF	0	SF	0
Handrail	0.25	320.0	LF	0	LF	0
Stair treads	0.85	192.0	EA	0	EA	0

6.56.2   Boiler Casing Sheet 2

	Installation Man Hour					
	Historical			Estimate		
	MH	Unit	Quantity	Unit		BM
Erect casing—generating bank, boiler, and penthouse	62	19.0	ton	0	ton	**0**

6.57 Structural Steel and Boiler Casing-Equipment Installation Man Hours

Facility—Biomass Plant Major Equipment	Actual	Estimated
	MH	BM
Erect boiler structural steel	4,601	0
Erect penthouse steel	1,100	0
Install platform, grating, and handrail at El. 100′-0″ to 184′-0″	6,140	0
Install stair no. 1 and no. 2	1,094	0
Erect casing—generating bank, boiler, and penthouse	1,180	0
Structural steel and boiler casing-equipment installation man hours	**14,116**	**0**

6.58 Biomass Plant Major Equipment Man Hour Breakdown

Facility Class—Biomass Plant Major Equipment	Actual	Estimate		
	MH	BM	MW	MH
400,000 lb/h waste heat boiler-equipment installation man hours	23114	0		**0**
Stoker-equipment and piping installation man hours	2236	0		**0**
Economizer-equipment installation man hours	824	0		**0**
Precipitator-equipment installation man hours	2512	0		**0**
Mechanical collector-equipment installation man hours	984	0		**0**
CO and No_x removal system-equipment installation man hours	1606	0		**0**
Wet scrubber-equipment installation man hours	688	0		**0**
Tubular air heater modules-equipment installation man hours	1172	0		**0**
Stack-equipment installation man hours	220	0		**0**
Fuel feeding equipment-equipment installation man hours	2742	0		**0**
Bottom ash drag system-equipment installation man hours	1264	0		**0**
ID, OFA, FD, and spout air fan-equipment installation man hours	686	0	0	**0**
ID, OFA, FD, underfire, and overfire ductwork installation man hours	3840	0		**0**
Structural steel and boiler casing-equipment installation man hours	14116	0		**0**
Biomass plant project man hours	56004	0	0	**0**

Ethanol Plant Equipment

7.1 General Scope of Field Work Required for Biomass Handling and Bagasse Storage

Scope of Work-Field Erection

Conveyor from shredder/miller to bagasse storage
Bagasse storage unit
Bagasse conveyor from storage to pretreatment

7.2 Biomass Handling and Bagasse Storage Estimate Data

7.2.1 Conveyors and Alignment Sheet 1

Description	MH	Unit
Conveyor-belt enclosed with walkway	2.62	LF
Conveyor alignment	0.60	LF

7.3 Equipment Installation Estimate

7.3.1 Conveyors and Alignment Sheet 1

	Historical			Estimate				
	MH	Quantity	Unit	Quantity	Unit	BM	IW	MW
Conveyor-Belt Enclosed With Walkway								
Conveyor from shredder/miller to bagasse storage	2.62	160.0	LF	0	LF		0	
Bagasse storage conveyor	2.62	90.0	LF	0	LF		0	
Bagasse conveyor from storage to pretreatment	2.62	556.0	LF	0	LF		0	
Conveyor alignment	0.60	**806.0**	LF	0	LF			0

Industrial Piping and Equipment Estimating Manual. http://dx.doi.org/10.1016/B978-0-12-813946-2.00007-1

7.4 Biomass Handling and Bagasse Storage-Equipment Installation Man Hours

Facility—Ethanol Plant Major Equipment	Actual	Estimate			
	MH	BM	IW	MW	MH
Conveyor from shredder/miller to bagasse storage	420		0		0
Bagasse storage conveyor	236		0		0
Bagasse conveyor from storage to pretreatment	1457		0		0
Conveyor alignment	484			0	0
Biomass handling and bagasse storage installation man hours	**2596**	**0**	**0**	**0**	**0**

7.5 General Scope of Field Work Required for Pretreatment

Pretreatment
Scope of Work-Field Erection
Install Major Equipment

C5 stream storage tank filter-cartridge filter
Sealing water cooler
C5 stream cooler
Demin water cooler
C5 stream transfer pump; 30 hp
C5 recirculation pump; 100 hp
C5 stream spare pump; 50 hp
Cooling water booster pump; 5 hp
Sealing water pump #1; 7 hp
Sealing water pump #2; 7 hp
Sealing water booster pump #1; 5 hp
Sealing water booster pump #2; 5 hp
Sealing water recirculation pump; 3 hp
C5 stream feed pump; 40 hp
Shop tank-C5 stream storage tank
Recycle drum
Sealing water tank
C5 stream storage tank nozzle #1, #2, #3
Proesa package/C5 stream treatment and membrane package
Centrifuge
Water heater
Cyclone
Tank filter
Membranes
Transfer pump
Liquid pump
Permeate pump
Retentive pump
Valve
Screw conveyor

7.6 Pretreatment Estimate Data

7.6.1 Pumps, Filters, Pressure Vessels, and Conveyors Sheet 1

Description	MH	Unit
Pumps		
Motor horsepower 15–30	1.60	hp
Motor horsepower 76–100	1.00	hp
Motor horsepower less than 15	16.00	EA
Motor horsepower 31–50	1.30	hp
Transfer, liquid, permeate, and retentive pumps	20.00	EA
Filters, coolers, tanks, and drums	24.00	EA
Filter	24.00	EA
Cooler up to 2000 lbs	40.00	ton
Cooler greater than 2001 lbs	20.00	ton
Tanks and drums	14.50	ton
Membranes	10.00	EA
Cyclone (knocked down) weight 0–40,000 lbs	32.00	ton
Centrifuge	160.00	EA
Water heater	20.00	EA
Pressure Vessels		
75,000 lb	1.60	ton
75,001 lb > weight < 125,000 lb	1.50	ton
125,001 lb > weight <= 200,000 lb	1.40	ton
200,001 lb > weight <= 300,000 lb	1.30	ton
Valve	5.43	ton
Conveyors-Stainless Steel Screw		
Screw conveyor; 24″ diameter	5.13	LF
Screw conveyor; 43″ diameter	6.75	LF
Screw conveyor; 66″ diameter	7.47	LF
Screw conveyor; 79″ diameter	7.69	LF
Screw conveyor; 81″ diameter	7.71	LF

7.7 Equipment Installation Estimate

7.7.1 Pumps, Filters, Pressure Vessels Sheet 1

		Historical		Estimate				
	MH	Quantity	Unit	Quantity	Unit	BM	IW	MW
Pumps						0	0	0
C5 stream transfer pump; 30 hp	1.60	30.0	hp	0	hp			0
C5 recirculation pump; 100 hp	1.00	100.0	hp	0	hp			0
C5 stream spare pump; 50 hp	1.20	50.0	hp	0	hp			0
Cooling water booster pump; 5 hp	16.00	1.0	EA	0	EA			0
Sealing water pump #1; 7 hp	16.00	1.0	EA	0	EA			0
Sealing water pump #2; 7 hp	16.00	1.0	EA	0	EA			0
Sealing water booster pump #1; 5 hp	16.00	1.0	EA	0	EA			0
Sealing water booster pump #2; 5 hp	16.00	1.0	EA	0	EA			0
Sealing water recirculation pump; 3 hp	16.00	1.0	EA	0	EA			0
C5 stream feed pump; 40 hp	1.30	40.0	hp	0	hp			0
Transfer pump	20.00	1.0	EA	0	EA			0
Liquid pump	20.00	1.0	EA	0	EA			0
Permeate pump	20.00	1.0	EA	0	EA			0
Retentive pump	20.00	1.0	EA	0	EA			0
Filters, coolers, tanks, and drums						0	0	0
C5 stream storage tank filter-cartridge filter	24.00	1.0	EA	0	EA			0
Sealing water cooler	40.00	0.5	ton	0	ton	0		
C5 stream cooler	20.00	1.8	ton	0	ton	0		
Demin water cooler	20.00	2.0	ton	0	ton	0		
Shop tank-C5 stream storage tank	14.50	10.2	ton	0	ton	0		
Recycle drum	14.50	3.4	ton	0	ton	0		
Sealing water tank	14.50	0.9	ton	0	ton	0		
Tank filter	24.00	4.0	EA	0	EA			0
Membranes	10.00	8.0	EA	0	EA			0
Centrifuge	160.00	1.0	EA	0	EA	0	0	0
Water heater	20.00	1.0	EA	0	EA	0	0	0
Cyclone	32.00	5.5	ton	0	ton	0	0	0
Cyclone	32.00	5.2	ton	0	ton	0	0	0
Vessels						0	0	0
Vessel diameter 177″ × 1299″; 198380#	1.40	99.2	ton	0	ton			0
Vessel diameter 87″ × 346″; 130038#	1.40	65.0	ton	0	ton			0
Vessel Wt = 28285#	1.60	14.1	ton	0	ton			0
Valve; Wt = 22040# (3 EA)	5.43	33.1	ton	0	ton			0

7.7.2   Conveyors and Alignment Sheet 2

	Historical			Estimate				
	MH	**Quantity**	**Unit**	**Quantity**	**Unit**	**BM**	**IW**	**MW**
Conveyors-stainless steel screw						**0**	**0**	**0**
Screw conveyor; 24″ diameter × 500″; 1180 H	5.13	41.7	LF	0	LF			0
Screw conveyor; 43″ diameter × 220″; 148 hp	6.75	18.3	LF	0	LF			0
Screw conveyor; 79″ diameter × 287″; 60 hp	7.69	23.9	LF	0	LF			0
Screw conveyor; 79″ diameter × 287″; 60 hp	7.69	23.9	LF	0	LF			0
Screw conveyor; 81″ diameter × 472″; 2146 hp	7.71	39.5	LF	0	LF			0
Screw conveyor; 79″ diameter × 125″; 30 hp	7.69	10.4	LF	0	LF			0
Screw conveyor; 66″ diameter × 276″; 101 hp	7.47	23.0	LF	0	LF			0
Conveyor alignment	1.42	**157.8**	LF	0	LF			0

7.8   Pretreatment-Equipment Installation Man Hours

Facility—Ethanol Plant Major Equipment	Actual	Estimate			
	MH	BM	IW	MW	MH
Pumps	436	0	0	0	0
Filters, coolers, tanks, and drums	504	0	0	0	0
Centrifuge	160	0	0	0	0
Water heater	20	0	0	0	0
Cyclone 5.5 ton	176	0	0	0	0
Cyclone 5.2 ton	166	0	0	0	0
Vessels	432	0	0	0	0
Conveyors-stainless steel screw	1486	0	0	0	0
Pretreatment-equipment installation man hours	**3380**	**0**	**0**	**0**	**0**

7.9   General Scope of Field Work Required for Viscosity Reduction

Scope of Work-Field Erection
Install Major Equipment

VR tank agitator
Hydrolysis tank agitator
Viscosity reduction tank filter-cartridge filter
Hydrolysis tank filter-cartridge filter
Viscosity reduction liquid cooler #1 and #2
Mash cooler #1, #2, and #3
Pretreated biomass pump; 125 hp
Viscosity reduction pump; 125 hp
Hydrolysis tank pump; 200 hp
Viscosity reduction tank spray nozzle #1 to #6
Hydrolysis tank spray nozzle #1 to #3

7.10　Viscosity Reduction Estimate Data

7.10.1　Pumps, Filter, Tanks, and Drums Sheet 1

Description	MH	Unit
Pumps		
Motor horsepower 15–30	1.60	hp
Motor horsepower 76–100	1.00	hp
Motor horsepower less than 15	16.00	EA
Motor horsepower 31–50	1.30	hp
Motor horsepower 101–125	0.90	hp
Motor horsepower 126–200	0.80	hp
Filters, Coolers, Tanks, and Drums		
Filter	24.00	EA
Cooler greater than 2001 lbs	20.00	ton
Tanks and drums	14.50	ton
Tank spray nozzle (1-1/2″ diameter)	4.00	EA
Tank agitator	16.00	ton

7.11　Equipment Installation Estimate

7.11.1　Pumps, Filter, Tanks, and Drums Sheet 1

Facility Class—Ethanol Plant Major Equipment	Historical			Estimate				
	MH	Quantity	Unit	Quantity	Unit	BM	PF	MW
Pumps						0	0	0
Pretreated biomass pump; 125 hp	0.90	125.0	hp	0	hp		0	0
Viscosity reduction pump; 125 hp	0.90	125.0	hp	0	hp		0	0
Hydrolysis tank pump; 200 hp	0.80	200.0	hp	0	hp		0	0
Filters, coolers, tanks, and drums						0	0	0
Viscosity reduction tank spray nozzle #1 to #6	4.00	6.0	EA	0	EA	0		
Hydrolysis tank spray nozzle #1 to #3	4.00	6.0	EA	0	EA	0		
Viscosity reduction tank filter-cartridge filter	24.00	1.0	EA	0	EA	0		
Hydrolysis tank filter-cartridge filter	24.00	1.0	EA	0	EA	0		
Viscosity reduction liquid cooler #1 and #2	20.00	6.0	ton	0	ton	0		
Mash cooler #1, #2, and #3	20.00	14.1	ton	0	ton	0		
Tank agitator						0	0	0
VR tank agitator	16.00	14.0	ton	0	ton	0		
Hydrolysis tank agitator	16.00	16.0	ton	0	ton	0		

7.12   Viscosity Reduction-Equipment Installation Man Hours

Facility—Ethanol Plant Major Equipment	Actual	Estimate			
	MH	BM	PF	MW	MH
Pumps	385	0	0	0	**0**
Filters, coolers, tanks, and drums	496	0	0	0	**0**
Tank agitator	479	0	0	0	**0**
Viscosity reduction-equipment installation man hours	**1360**	**0**	**0**	**0**	**0**

Fermentation

7.13   General Scope of Field Work Required for Fermentation

Scope of Work-Field Erection

Fermented #1 to #6 agitator
Yeast mix tank agitator
Yeast propagation tank #1 and #2 agitator
Beer well agitator
CO_2 scrubber
Yeast propagation tank #1 filter
Yeast mix tank filter
Sterile air prefilter
Sterile air filter
Low-pressure CIP separator filter
Fermenter #1 to #6 cooler
Yeast propagation #1 and #2 cooler
High-pressure CIP supply pump; 125 hp
Low-pressure CIP supply pump; 125 hp
Lignin filter cloth washing pump #1 and #2; 15 hp
Primary scrubber pump; 5 hp
Yeast mix tank
Thermal oxidizer package; 27′ × 11′
Modules
Stack
Duct/W supports
CEMS
Dampers/Exp Jt
Scrubber
Ethanol loading flare; 5MMBTU/h flare
Fermenter #1 to #6 spray nozzle
Yeast mix tank spray ball
Yeast propagation #1 CIP nozzle #1 and #2
Yeast propagation #2 CIP nozzle #1 and #2;
Bar well spray nozzle #1, #2, #3
Sparger tank - 3202
Sparger tank - 3203

7.14 Fermentation Estimate Data

7.14.1 Pumps, Filter, Vessels, and Modules Sheet 1

Description	MH	Unit
Pumps		
Motor horsepower 15–30	1.60	hp
Motor horsepower 76–100	1.00	hp
Motor horsepower less than 15	16.00	EA
Motor horsepower 31–50	1.30	hp
Motor horsepower 101–125	0.90	hp
Motor horsepower 126–200	0.80	hp
Filters, Coolers, Tanks, and Drums		
Filter	24.00	EA
Cooler greater than 2001 lbs	20.00	ton
Tanks and drums	14.50	ton
Tank spray nozzle (1-1/2″ diameter)	4.00	EA
Tank agitator	16.00	ton
CO$_2$ scrubber	25.00	ton
Vessel—tray packed (67″ diameter)	20.00	EA
Thermal Oxidizer Package; 27′ × 11′		
Modules		
75,000 lb	2.50	ton
Flues and ducts, including supports, hoppers, doors, expansion joints, and dampers	40.00	ton
Stack (60″ diameter)	1.80	LF
CEMS	48.00	EA
Ethanol loading flare; 5MMBTU/h flare	320	EA

7.15 Equipment Installation Estimate

7.15.1 Pumps, Filter, Vessels, and Modules Sheet 1

		Historical		Estimate				
	MH	Quantity	Unit	Quantity	Unit	BM	PF	MW
Pumps						**0**	**0**	**0**
High-pressure CIP	0.90	125.0	hp	0	hp		0	0
supply pump; 125 hp								
Low-pressure CIP	0.90	125.0	hp	0	hp		0	0
supply pump; 125 hp								
Lignin filter cloth washing	1.60	30.0	hp	0	hp		0	0
pump #1 and #2; 15 hp								
Primary scrubber pump; 5 hp	16.00	1.0	EA	0	EA		0	0
Filters, coolers, tanks, and						**0**	**0**	**0**
drums								
Yeast propagation tank #1 filter	24.00	2.0	EA	0	EA	0		
Yeast mix tank filter	24.00	1.0	EA	0	EA	0		
Sterile air prefilter	24.00	1.0	EA	0	EA	0		
Sterile air filter	24.00	1.0	EA	0	EA	0		
Low-pressure CIP separator filter	24.00	1.0	EA	0	EA	0		
Fermenter #1 to #6 cooler	20.00	24.0	ton	0	ton	0		
Yeast propagation #1 and #2	20.00	4.4	ton	0	ton	0		
cooler								
Yeast mix tank	14.50	0.6	ton	0	ton	0		
Sparger tank—3202 and 3203	14.50	2.5	ton	0	ton	0		
Fermenter #1 to #6 spray nozzle	4.00	18.0	EA	0	EA	0		
Yeast mix tank spray ball	4.00	1.0	EA	0	EA	0		
Yeast propagation #1 CIP nozzle	4.00	2.0	EA	0	EA	0		
#1 and #2								
Yeast propagation #2 CIP nozzle	4.00	2.0	EA	0	EA	0		
#1 and #2								
Bar well spray nozzle #1, #2, #3	4.00	2.0	EA	0	EA	0		
Tank agitator						**0**	**0**	**0**
Fermenter #1 to #6 agitator	16.00	33.1	ton	0	ton	0		
Yeast mix tank agitator	16.00	0.3	ton	0	ton	0		
Yeast propagation tank #1 and	16.00	5.6	ton	0	ton	0		
#2 agitator								
Beer well agitator	16.00	15.4	ton	0	ton	0		
CO_2 Scrubber;	25.00	4.2	ton	0	ton	**0**	**0**	**0**
67′ dia × 287′								
Vessel—tray packed (67′	20.00	1.0	EA	0	EA	**0**	**0**	**0**
diameter)								
Thermal oxidizer pack-						**0**	**0**	**0**
age; 27′ × 11′								
Modules	2.50	16.0	ton	0	ton	0		
Stack	1.80	100.0	LF	0	LF	0		
Duct/W supports	40.00	3.0	ton	0	ton	0		
CEMS	48.00	1.0	EA	0	EA	0		
Dampers/expansion joint	40.00	1.5	ton	0	ton	0		
Scrubber	25.00	2.4	ton	0	ton	0		
Ethanol loading flare;	320	1.0	EA	0	EA	**0**	**0**	**0**
5MMBTU/h flare								

7.16   Fermentation-Equipment Installation Man Hours

Facility—Ethanol Plant Major Equipment	Actual	Estimate			
	MH	BM	PF	MW	MH
Pumps	289	0	0	0	0
Filters, coolers, tanks, and drums	857	0	0	0	0
Tank agitator	871	0	0	0	0
CO_2 scrubber; 67″ dia × 287″	105	0	0	0	0
Vessel—tray packed (67″ diameter)	20	0	0	0	0
Thermal oxidizer package; 27′ × 11″	508	0	0	0	0
Ethanol loading flare; 5MMBTU/h flare	320	0	0	0	0
Fermentation-equipment installation man hours	**2970**	**0**	**0**	**0**	**0**

7.17   General Scope of Field Work Required for Distillation and Dehydration

Distillation and dehydration
Scope of Work-Field Erection
Beer Column; 169″ dia × 780″; Trays 22 EA; Wt = 125,995#

Unload, handle, haul up to 2000′, rig, set and align, make up Foundation AB
Install platforms and ladders
Remove and replace manway cover (24″ 300# Removable-Davit)
Install double downflow valve trays (22 trays)
Install demisting pads (single grid-support, pad, grid-top)
Vortex breaker
Packing (pall rings)

Rectifier Column; 71″ dia × 1226″; Trays 57 EA; Wt = 61,973#

Unload, handle, haul up to 2000′, rig, set and align, make up Foundation AB
Install platforms and ladders
Remove and replace manway cover (24″ 300# Removable-Davit)
Install double downflow valve trays (57 trays)
Install demisting pads (single grid-support, pad, grid-top)
Vortex breaker
Packing (pall rings)
Regeneration filter #1 and #2; bag filter; 10″ dia × 28″
Product filter #1 and #2; bag filter; 11″ dia × 42″
Beer column reboiler #1 and #2; 3725 SF
Vent condenser; 145 SF; Wt = 1348#
Excess stream condenser; 1735 SF; Wt = 1151#
Rectifier feed preheater #1; 118 SF; Wt = 545#
Rectifier column reboiler; 2359 SF; Wt = 12,871#
Heavy alcohols heat exchanger; 14″ dia × 441″
Overhead condenser #1, #2 and #3; 188,283,744 SF; Wt = 1543#, 2865#, 4370#
Molecular sieve superheater; 28 SF; 496#
Regeneration vacuum condenser; 28 SF
Ethanol condenser #1; 284 SF; Wt = 3152#
Ethanol cooler; 85 SF; Wt = 441#
Regeneration cooler; 128 SF; Wt = 323#
Ethanol condenser #3; 355 SF; Wt = 2327#
Product cooler; 85SF; Wt = 470#
Beer column pump #1 and #2; 100 hp
Rectifier feed pump; 15 hp
LP steam condensate pump #1 and #2; 10 hp
Scrubber water pump; 25 hp
Beer reboiler pump; 3 hp
Rectifier reflux pump; 15 hp
Water pump; 3 hp
Heavy alcohols pump; 3 hp

7.18   General Scope of Field Work Required for Distillation and Dehydration

Scope of Work-Field Erection

Product pump; 5 hp
Regeneration pump; 40 hp
Rectifier feed tank; 60″ dia × 168″; Wt = 6381#
LP steam condenser receiver; 60″ dia × 174″; Wt = 8180#
LP steam scrubber; 55″ dia × 377″; Wt = 15548#
Reboiler vapor separator; 120″ dia × 204″; Wt = 37623#
Rectifier reflux receiver; 54″ dia × 138″; Wt = 4093#
Heavy alcohols separator; 63″ dia × 106″; Wt = 3637#
Heavy alcohols tank; 118″ dia × 157″; Wt = 11021#
Molecular sieve bed #1 and #2; 102″ dia × 252″; Wt = 30180
Product receiver; 54″ dia × 114″; Wt = 5188#
Regeneration receiver; 60″ dia × 120″; Wt = 8486#

Beer Column Vacuum System Package

Heat exchanger
Vacuum pump
Vessel
Aux pump

Regeneration Receiver Vacuum System Package

Heat exchanger
Liquid ring vacuum pump
Aux pump

Exhaust Steam Blower
Beer Column Trays Internal (Included in Beer Column Estimate)
Rectifier Column Trays Internal (Included in Rectifier Column Estimate)
MDL Sieve Bid Internals
Beer Column Steam Sparger

7.19   Distillation and Dehydration Estimate Data

7.19.1   Pumps, Condensers, Vessels Sheet 1

Description	MH	Unit
Pumps		
Motor horsepower 15–30	1.60	hp
Motor horsepower 76–100	1.00	hp
Motor horsepower less than 15	16.00	EA
Motor horsepower 31–50	1.30	hp
Motor horsepower 101–125	0.90	hp
Motor horsepower 126–200	0.80	hp
Motor horsepower less than 15	16.00	EA
Filters, Coolers, Tanks, and Drums		
Regeneration filter #1 and #2; bag filter	24.00	EA
Product filter #1 and #2; bag filter	24.00	EA
Cooler-ethanol regeneration, product	40.00	ton
Heavy alcohols tank and rectifier feed tank	14.50	ton
Condenser		
Condenser s-vent, steam, overhead, vacuum, ethanol, and LP steam	28.00	ton
Vessels		
Receiver-product, regeneration, reboiler, heavy alcohols	24.00	ton
Reboiler vapor separator	9.55	ton
LP steam scrubber	14.70	ton
Rectifier feed preheater #1	40.00	ton
Rectifier column reboiler, heavy alcohols exchanger, molecular sieve bed	14.70	ton
Rectifier feed preheater #1 and molecular sieve superheater	40.00	ton
Beer column reboiler #1 and #2	40.00	EA

7.19.2   Beer Column, Rectifier Column, Vacuum System Package, and Trays Sheet 2

Description	MH	Unit
Beer Column; 169″ dia × 780″; Trays 22 EA; Wt = 125995#		
Unload, handle, haul up to 2000′, rig, set and align, make up foundation AB	8.57	ton
Install platforms and ladders	80.00	LOT
Remove and replace manway cover (24″ 300# removable davit)	40.00	EA
Install double downflow valve trays (22 trays)	40.00	EA
Install demisting pads (single grid-support, pad, grid-top)	100.00	EA
Vortex breaker	44.00	EA
Packing (pall rings)	36.00	LOT
Rectifier Column; 71″ dia × 1226″; Trays 57 EA; Wt = 61973#		
Unload, handle, haul up to 2000′, rig, set and align, make up foundation AB	10.32	ton
Install platforms and ladders	60.00	LOT
Remove and replace manway cover (24″ 300# removable-davit)	40.00	EA
Install double downflow valve trays (57 trays)	20.00	EA
Install demisting pads (single grid-support, pad, grid-top)	40.00	EA
Vortex breaker	20.00	EA
Packing (pall rings)	20.00	LOT
Beer Column Vacuum System Package: Regeneration Receiver Vacuum Package		
Heat exchanger	48.00	EA
Vacuum pump	40.00	EA
Vessel	60.00	EA
Aux pump	60.00	EA
Liquid ring vacuum pump	40.00	EA
Exhaust steam blower	35.00	EA
Beer Column Trays Internal (Included in Beer Column Estimate)		
Rectifier Column Trays Internal (Included in Rectifier Column Estimate)		
MDL sieve bid internals	40.00	EA
Beer column steam sparger	24.00	EA

7.20 Equipment Installation Estimate

7.20.1 Pumps, Condensers, Vessels Sheet 1

	Historical			Estimate				
	MH	Quantity	Unit	Quantity	Unit	BM	PF	MW
Pumps						0	0	0
Beer column pump #1 and #2; 100 hp	1.00	200.0	hp	0	hp		0	0
Rectifier feed pump; 15 hp	1.60	15.0	hp	0	hp		0	0
LP steam condensate pump #1 and #2; 10 hp	16.00	2.0	EA	0	EA		0	0
Scrubber water pump; 25 hp	1.60	25.0	hp	0	hp		0	0
Beer reboiler pump; 3 hp	16.00	1.0	EA	0	EA		0	0
Rectifier reflux pump; 15 hp	1.60	15.0	hp	0	hp		0	0
Water pump; 3 hp	16.00	1.0	EA	0	EA		0	0
Heavy alcohols pump; 3 hp	16.00	1.0	EA	0	EA		0	0
Product pump; 5 hp	16.00	1.0	EA	0	EA		0	0
Regeneration pump; 40 hp	1.30	40.0	hp	0	hp		0	0
Filters, coolers, tanks, and drums						0	0	0
Regeneration filter #1 and #2; bag filter	24.00	2.0	EA	0	EA	0		
Product filter #1 and #2; bag filter	24.00	2.0	EA	0	EA	0		
Ethanol cooler	40.00	0.2	ton	0	ton	0		
Regeneration cooler	40.00	0.2	ton	0	ton	0		
Product cooler	40.00	0.2	ton	0	ton	0		
Heavy alcohols tank	14.50	5.5	ton	0	ton	0		
Rectifier feed tank	14.50	3.2	ton	0	ton	0		
Condenser						0	0	0
Vent condenser	28.00	0.7	ton	0	ton	0		
Excess stream condenser	28.00	0.6	ton	0	ton	0		
Overhead condenser #1, #2, and #3	28.00	4.4	ton	0	ton	0		
Regeneration vacuum condenser	28.00	0.3	ton	0	ton	0		
Ethanol condenser #1	28.00	1.6	ton	0	ton	0		
Ethanol condenser #3	28.00	1.2	ton	0	ton	0		
LP steam condenser receiver	28.00	4.1	ton	0	ton	0		
Vessels						0	0	0
Product receiver	24.00	2.6	ton	0	ton	0		
Regeneration receiver	24.00	4.2	ton	0	ton	0		
Rectifier reflux receiver	24.00	2.0	ton	0	ton	0		
Reboiler vapor separator	9.55	18.8	ton	0	ton	0		
Heavy alcohols separator	24.00	1.8	ton	0	ton	0		
LP steam scrubber	14.70	7.8	ton	0	ton	0		
Rectifier feed preheater #1; 118 SF; Wt = 545#	40.00	0.3	ton	0	ton	0		
Rectifier column reboiler; 2359 SF; Wt = 12871#	14.70	6.4	ton	0	ton	0		
Heavy alcohols heat exchanger; 14″ dia × 441″	14.70	2.5	ton	0	ton	0		
Molecular sieve superheater; 28 SF; 496#	40.00	0.2	ton	0	ton	0		
Molecular sieve bed #1 and #2	14.70	15.1	ton	0	ton	**0**	**0**	**0**
Beer column reboiler #1 and #2	40.00	2.0	EA	0	EA	**0**	**0**	**0**

7.20.2   Beer Column, Rectifier Column, Vacuum System Package, and Trays Sheet 2

	Historical			Estimate				
	MH	Quantity	Unit	Quantity	Unit	BM	PF	MW
Beer column; 169″ dia × 780″; trays 22 EA						0		
Rig, set and align, make up foundation AB	8.57	63.0	ton	0	ton	0		
Install platforms and ladders	80.00	1.0	LOT	0	LOT	0		
Remove and replace manway cover	40.00	1.0	EA	0	EA	0		
Install double downflow valve trays (22 trays)	40.00	22	EA	0	EA	0		
Install demisting pads	100.00	1.0	EA	0	EA	0		
Vortex breaker	44.00	1.0	EA	0	EA	0		
Packing (pall rings)	36.00	1.0	LOT	0	LOT	0		
Rectifier column; 71″ dia × 1226″; trays 57 EA						0		
Rig, set and align, make up foundation AB	10.32	31.0	ton	0	ton	0		
Install platforms and ladders	60.00	1.0	LOT	0	LOT	0		
Remove and replace manway cover	40.00	1.0	EA	0	EA	0		
Install double downflow valve trays (57 trays)	20.00	57	EA	0	EA	0		
Install demisting pads	40.00	1.0	EA	0	EA	0		
Vortex breaker	20.00	1.0	EA	0	EA	0		
Packing (pall rings)	20.00	1.0	LOT	0	LOT	0		
Beer column vacuum system package						0	0	0
Heat exchanger	48.00	1.0	EA	0	EA	0		
Vacuum pump	40.00	1.0	EA	0	EA			0
Vessel	60.00	1.0	EA	0	EA	0		
Aux pump	60.00	1.0	EA	0	EA			0
Regeneration receiver vacuum system package						0	0	0
Heat exchanger	48.00	1.0	EA	0	EA	0		
Liquid ring vacuum pump	40.00	1.0	EA	0	EA			0
Aux pump	60.00	1.0	EA	0	EA			0
Exhaust steam blower	35.00	1.0	EA	0	EA	0	0	0
Beer Column Trays Internal (In Estimate)								
Rectifier column trays internal (in estimate)				0				
MDL sieve bid internals	40.00	1.0	EA	0	EA	0	0	0
Beer column steam sparger	24.00	1.0	EA	0	EA	0	0	0

7.21 Distillation and Dehydration-Equipment Installation Man Hours

Facility—Ethanol Plant Major Equipment	Actual	Estimate			
	MH	BM	PF	MW	MH
Pumps	436	0	0	0	0
Filters, coolers, tanks, and drums	247	0	0	0	0
Condenser	358	0	0	0	0
Vessels	703	0	0	0	0
Molecular sieve bed #1 and #2	222	0	0	0	0
Beer column reboiler #1 and #2	80	0	0	0	0
Beer column; 169″ dia×780″; trays 22 EA	1720	0	0	0	0
Rectifier column; 71″ dia×1226″; trays 57 EA	1640	0			0
Beer column vacuum system package	208	0	0	0	0
Regeneration receiver vacuum system package	148	0	0	0	0
Exhaust steam blower	35	0	0	0	0
MDL sieve bid internals	40	0	0	0	0
Beer column steam sparger	24	0	0	0	0
Distillation and dehydration-equipment installation man-hours	**5860**	**0**	**0**	**0**	**0**

Ethanol Storage and Loading.

7.22 General Scope of Field Work Required for Ethanol Storage and Loading

Scope of Work-Field Erection

Ethanol filter #1 and #2; 30″ dia×49″
Ethanol transfer pump #1, #2 and #3; 20 hp
Off spec pump; 10 hp
Denetarant pump; 3 hp
Denetarant unloading pump; 7 hp
Product loadout pump #1 and #2; 15 hp
Fiscal measurement rail package (Scale); 1 unit
Prover skid, meters, panels, ample system w/cabinets, pumps, IC piping, receiver, scale
Fiscal measurement package; 2 units
Prover skid, meters, panels, sample system w/cabinets, pumps, IC piping, receiver
Truck ethanol loading package; 2 units
Skid, piping, valves, loading arms/w swivel, rack monitors

7.23   Ethanol Storage and Loading Estimate Data

7.23.1   Pumps, Filters, Coolers, Tanks, Fiscal Measurements and Truck Loading Package Sheet 1

Description	MH	Unit
Pumps		
Ethanol transfer pump #1, #2, and #3; 20 hp	1.60	hp
Off spec pump; 10 hp	16.00	EA
Denetarant pump; 3 hp	16.00	EA
Denetarant unloading pump; 7 hp	16.00	EA
Product loadout pump#1 and #2; 15 hp	1.60	hp
Filters, Coolers, Tanks, and Drums		
Ethanol filter #1 and #2; 30″ dia × 49″	24.00	EA
Fiscal Measurement Rail Package (Scale); 1 Unit		
Prover skid, meters, panels, sample system w/cabinets, pumps, IC piping, receiver, scale	180.00	Unit
Fiscal Measurement Package; 2 Units		
Prover skid, meters, panels, sample system w/cabinets, pumps, IC piping, receiver	160.00	Unit
Truck Ethanol Loading Package; 2 Units		
Skid, piping, valves, loading arms/w swivel, rack monitors	140.00	Unit

7.24   Equipment Installation Estimate

7.24.1   Pumps, Filters, Coolers, Tanks, Fiscal Measurements, and Truck Loading Package Sheet 1

	Historical			Estimate				
	MH	Quantity	Unit	Quantity	Unit	BM	PF	MW
Pumps						0	0	0
Ethanol transfer pump #1, #2, and #3; 20 hp	1.60	60.0	hp	0	hp		0	0
Off spec pump; 10 hp	16.00	1.0	EA	0	EA		0	0
Denetarant pump; 3 hp	16.00	1.0	EA	0	EA		0	0
Denetarant unloading pump; 7 hp	16.00	1.0	EA	0	EA		0	0
Product loadout pump #1 and #2; 15 hp	1.60	30.0	hp	0	hp		0	0
Filters, coolers, tanks, and drums						0	0	0
Ethanol filter #1 and #2; 30″ dia × 49″	24.00	2.0	EA	0	EA	0		
Fiscal Measurement Rail Package (Scale); 1 Unit								
Prover skid, pumps, IC piping, receiver, scale	180.00	1.0	Unit	0	Unit	0	0	0
Fiscal Measurement Package; 2 Units								
Prover skid, pumps, IC piping, receiver, scale	160.00	2.0	Unit	0	Unit	0	0	0
Truck Ethanol Loading Package; 2 Units								
Skid, piping, valves, loading arms/w swivel	140.00	2.0	Unit	0	Unit	0	0	0

7.25 Ethanol Storage and Loading-Equipment Installation Man Hours

	Actual	Estimate			
Facility—Ethanol Plant Major Equipment	MH	BM	PF	MW	MH
Pumps	192	0	0	0	**0**
Filters, coolers, tanks, and drums	48	0	0	0	**0**
Fiscal measurement rail package (scale); 1 unit	540	0	0	0	**0**
Fiscal measurement package; 2 units	960	0	0	0	**0**
Truck ethanol loading package; 2 units	840	0	0	0	**0**
Ethanol storage and loading-equipment installation man-hours	**2580**	**0**	**0**	**0**	**0**

7.26 General Scope of Field Work Required for Chemical Storage

Scope of Work-Field Erection

Urea tank agitator; 150″ dia×315″; Wt=4298#
Sulfuric acid filter; 4″dia×22″
Enzyme filter #1 and #2; 4″ dia×22″
Enzyme cooler #1 and #2; 108 SF; Wt=3306#
Caustic transfer pump; 15 hp
Sulfuric acid pump; 1 hp
Enzyme pump #1 and #2; 15 hp
Metering pump to viscosity reduction #1 and #2; 5 hp
Urea pump; 1 hp
Urea unloading pump; 15 hp
Potassium hydroxide pump; 1 hp
Antifoam pump; 1 hp
Sulfuric acid unloading pump; 15 hp
Caustic pump; 1 hp
Potassium hydroxide unloading pump; 10 hp
Antifoam unloading pump; 15 hp
Pretreatment area sump pump; 15 hp
Fermentation area sump pump; 15 hp
Main area sump pump; 15 hp
Solid separation area sump pump; 15 hp
Tank farm area sump pump; 15 hp
Fire water AC pump #1; 60 hp
Fire water AC pump #2; 250 hp
Caustic tank; 138″ dia×236″; Wt=11,796#
Sulfuric acid tank; 138″ dia×236″; Wt=9014#
Enzyme tank #1 and #2; 138″ dia×236″; Wt=9015#
Potassium hydroxide tank; 149″ dia×314″; Wt=12,270#
Urea tank; 149″×314″; Wt=12,270#
Antifoam tank; 138″ dia×236″; Wt=40,56#
Enzyme tank #1 and #2 spray nozzle #1, #2, #3; 0.375″ NPT; Wt=0.7#

7.27 Chemical Storage Estimate Data

7.27.1 Pumps, Filters, Coolers, Tanks, and Drums Sheet 1

Description	MH	Unit
Pumps		
Motor horsepower 15–30	1.60	hp
Motor horsepower 76–100	1.00	hp
Motor horsepower less than 15	16.00	EA
Motor horsepower 31–50	1.30	hp
Motor horsepower 101–125	0.90	hp
Motor horsepower 126–200	0.80	hp
Motor horsepower less than 15	16.00	EA
Filters, Coolers, Tanks, and Drums		
Urea tank agitator; 150″ dia×315″; Wt=4298#	16.00	ton
Sulfuric acid filter; 4″dia×22″	24.00	EA
Enzyme filter #1 and #2; 4″ dia×22″	24.00	EA
Enzyme cooler #1 and #2; 108 SF; Wt=3306#	40.00	ton
Caustic tank; 138″ dia×236″; Wt=11796#	14.50	ton
Sulfuric acid tank; 138″ dia×236″; Wt=9014#	14.50	ton
Enzyme tank #1 and #2; 138″ dia×236″; Wt=9015#	14.50	ton
Potassium hydroxide tank; 149″ dia×314″; Wt=12270#	14.50	ton
Urea tank; 149″×314″; Wt=12270#	14.50	ton
Antifoam tank; 138″ dia×236″; Wt=4056#	14.50	ton
Enzyme tank #1 and #2 spay nozzle #1, #2, #3 and #1, #2, #3; 0.375″ NPT; Wt=0.7#	4.00	EA

7.28   Equipment Installation Estimate

7.28.1   Pumps, Filters, Coolers, Tanks, and Drums Sheet 1

	Historical			Estimate				
	MH	Quantity	Unit	Quantity	Unit	BM	PF	MW
Pumps						**0**	**0**	**0**
Caustic transfer pump; 15 hp	1.60	15.0	hp	0	hp		0	0
Sulfuric acid pump; 1 hp	16.00	1.0	EA	0	EA		0	0
Enzyme pump #1 and #2; 15 hp	1.60	30.0	hp	0	hp		0	0
Metering pump	16.00	2.0	EA	0	EA		0	0
Urea pump; 1 hp	16.00	1.0	EA	0	EA		0	0
Urea unloading pump; 15 hp	1.60	15.0	hp	0	hp		0	0
Potassium hydroxide pump; 1 hp	16.00	1.0	EA	0	EA		0	0
Antifoam pump; 1 hp	16.00	1.0	EA	0	EA		0	0
Sulfuric acid unloading pump; 15 hp	1.60	15.0	hp	0	hp		0	0
Caustic pump; 1 hp	16.00	1.0	EA	0	EA		0	0
Potassium hydroxide unloading pump; 10 hp	16.00	1.0	EA	0	EA		0	0
Antifoam unloading pump; 15 hp	1.60	15.0	hp	0	hp		0	0
Fermentation area sump pump; 15 hp	1.60	15.0	hp	0	hp		0	0
Main area sump pump; 15 hp	1.60	15.0	hp	0	hp		0	0
Solid separation area sump pump; 15 hp	1.60	15.0	hp	0	hp		0	0
Tank farm area sump pump; 15 hp	1.60	15.0	hp	0	hp		0	0
Fire water AC pump #1; 60 hp	1.60	60.0	hp	0	hp		0	0
Fire water AC pump #2; 250 hp	0.80	250.0	hp	0	hp		0	0
Filters, coolers, tanks, and drums						**0**	**0**	**0**
Urea tank agitator; 150″ dia×315″; Wt=4298#	16.00	2.1	ton	0	ton	0		
Sulfuric acid filter; 4″ dia×22″	24.00	1.0	EA	0	EA	0		
Enzyme filter #1 and #2; 4″ dia×22″	24.00	2.0	EA	0	EA	0		
Enzyme cooler #1 and #2; 108 SF; Wt=3306#	40.00	3.3	ton	0	ton	0		
Caustic tank; 138″ dia×236″; Wt=11796#	14.50	5.9	ton	0	ton	0		
Sulfuric acid tank; 138″ dia×236″; Wt=9014#	14.50	4.5	ton	0	ton	0		
Enzyme tank #1 and #2; 138″ dia×236″; Wt=9015#	14.50	4.5	ton	0	ton	0		
Potassium hydroxide tank; 149″ dia×314″	14.50	6.1	ton	0	ton	0		
Urea tank; 149″×314″; Wt=12270#	14.50	6.1	ton	0	ton	0		
Antifoam tank; 138″ dia×236″; Wt=4056#	14.50	2.0	ton	0	ton	0		
Enzyme tank #1 and #2 spray nozzle	4.00	8.0	EA	0	EA	0		

7.29 Chemical Storage-Equipment Installation Man Hours

	Actual	Estimate			
Facility—Ethanol Plant Major Equipment	MH	BM	PF	MW	MH
Pumps	688	0	0	0	0
Filters, coolers, tanks, and drums	694	0	0	0	0
Chemical storage-equipment installation man-hours	**1382**	**0**	**0**	**0**	**0**

Lignin Separation

7.30 General Scope of Field Work Required for Lignin Separation

Scope of Work-Field Erection

Lignin conveyor #1 to #6; 7′×50′; 7.5 hp

7.31 Equipment Installation Estimate

7.31.1 Conveyor Sheet 1

	Historical			Estimate				
	MH	Quantity	Unit	Quantity	Unit	BM	IW	MW
Lignin conveyor #1 to #6; 7′×50′; 7.5 hp	3.60	300	LF	0	LF	**0**	**0**	**0**

7.32 Lignin Separation-Equipment Installation Man Hours

	Actual	Estimate			
Facility—Ethanol Plant Major Equipment	MH	BM	IW	MW	MH
Lignin conveyor #1 to #6; 7′×50′; 7.5 hp	1080	0	0	0	0
Lignin separation-equipment installation man-hours	**1080**	**0**	**0**	**0**	**0**

7.33 General Scope of Field Work Required for Lignin Storage and Handling

Scope of Work-Field Erection

Slurry feed pump #1 to #6; 75 hp
Clarified stillage pump #1, #2; 150 hp
Clarified stillage transfer pump #1; 15 hp
Clarified stillage transfer pump #2; 100 hp
Drain stillage transfer pump; 5 hp
Stillage tank agitator
Lignin handling and storage package; 2 units w/conveyor
Lignin storage tank w/reclaim; 396″ dia × 336″; Wt = 9918#
Stillage tank spray nozzle #1, #2 and #3; 1″ NPT; Wt = 11.2#

7.34 Lignin Storage and Handling Estimate Data

7.34.1 Pumps, Filters, Coolers, Tanks, and Drums, Conveyor Sheet 1

Description	MH	Unit
Pumps		
Motor horsepower 15–30	1.60	hp
Motor horsepower 76–100	1.00	hp
Motor horsepower less than 15	16.00	EA
Motor horsepower 31–50	1.30	hp
Motor horsepower 101–125	0.90	hp
Motor horsepower 126–200	0.80	hp
Motor horsepower less than 15	16.00	EA
Filters, Coolers, Tanks, and Drums		
Stillage tank agitator	16.00	ton
Lignin storage tank w/reclaim; 396″ dia × 336″; Wt = 9918#	14.50	ton
Stillage tank spray nozzle #1, #2 and #3; 1″ NPT; Wt = 11.2#	4.00	EA
Lignin handling and storage package; 2 units w/conveyor	3.60	LF

7.35 Equipment Installation Estimate

7.35.1 Pumps, Filters, Coolers, Tanks, and Drums, Conveyor Sheet 1

	Historical			Estimate				
	MH	Quantity	Unit	Quantity	Unit	BM	PF	MW
Pumps						0	0	0
Slurry feed pump #1 to #6; 75 hp	1.00	450.0	hp	0	hp		0	0
Clarified stillage pump #1, #2; 150 hp	0.80	300.0	hp	0	hp		0	0
Clarified stillage transfer pump #1; 15 hp	1.60	25.0	hp	0	hp		0	0
Clarified stillage transfer pump #2; 100 hp	0.90	100.0	hp	0	hp		0	0
Drain stillage transfer pump; 5 hp	16.00	1.0	EA	0	EA		0	0
Filters, coolers, tanks, and drums						0	0	0
Stillage tank agitator	16.00	2.1	ton	0	ton	0		
Lignin storage tank w/ reclaim	14.50	5.0	ton	0	ton	0		
Stillage tank spray nozzle	4.00	3.0	EA	0	EA	0		
Lignin handling and storage package; 2 units w/conveyor	3.60	100.0	LF	0	LF	0	0	0

7.36 Lignin Storage and Handling-Equipment Installation Man Hours

	Actual	Estimate			
Facility—Ethanol Plant Major Equipment	MH	BM	PF	MW	MH
Pumps	836	0	0	0	**0**
Filters, coolers, tanks, and drums	118	0	0	0	**0**
Lignin handling and storage package; 2 units w/conveyor	360	0	0	0	**0**
Lignin storage and handling-equipment installation man hours	**1314**	**0**	**0**	**0**	**0**

7.37 General Scope of Field Work Required for Wastewater Treatment

Scope of Work-Field Erection

WW storage tank agitator; 472″ dia × 512″; Wt = 882#
Concentrated stillage tank agitator; 150″ dia × 315″; Wt = 22,040#
Sewage water pump; 3 hp
Storm water pump; 7 hp
WW upset treatment pump; 5 hp
WW treatment pump; 75 hp
Treated WW pump; 30 hp
Treated WW hp pump; 200 hp
Concentrated stillage pump; 5 hp
Process water UV sterilization package; 52″ × 21″ × 80″; Wt = 500#
Diesel generator set; 1000 hp
Evaporator package; 95′ × 40′ × 80′

7.38 Wastewater Treatment Estimate Data

7.38.1 Pumps, Tanks, Process Water Package, Generator, Evaporator Sheet 1

Description	MH	Unit
Pumps		
Motor horsepower 15–30	1.60	hp
Motor horsepower 76–100	1.00	hp
Motor horsepower less than 15	16.00	EA
Motor horsepower 31–50	1.30	hp
Motor horsepower 101–125	0.90	hp
Motor horsepower 126–200	0.80	hp
Motor horsepower less than 15	16.00	EA
Filters, Coolers, Tanks, and Drums		
WW storage tank agitator; 472″ dia × 512″; Wt = 882#	16.00	ton
Concentrated stillage tank agitator; 150″ dia × 315″; Wt = 22040#	16.00	ton
Process water UV sterilization package; 52″ × 21″ × 80″; Wt = 500#	60.00	EA
Diesel generator set; 1000 hp	120.00	EA
Evaporator package; 95′ × 40′ × 80′	320.00	EA

7.39 Equipment Installation Estimate

7.39.1 Pumps, Tanks, Process Water Package, Generator, Evaporator Sheet 1

	Historical			Estimate				
	MH	Quantity	Unit	Quantity	Unit	BM	PF	MW
Pumps						0	0	0
Sewage water pump; 3 hp	16.00	1.0	EA	0	EA		0	0
Storm water pump; 7 hp	16.00	1.0	EA	0	EA		0	0
WW upset treatment pump; 5 hp	16.00	1.0	EA	0	EA		0	0
WW treatment pump; 75 hp	1.00	75.0	hp	0	hp		0	0
Treated WW pump; 30 hp	1.60	30.0	hp	0	hp		0	0
Treated WW HP pump; 200 hp	0.80	200.0	hp	0	hp		0	0
Concentrated stillage pump; 5 hp	16.00	1.0	EA	0	EA		0	0
Filters, coolers, tanks, and drums						0	0	0
WW storage tank agitator; 472″ dia × 512″	16.00	0.4	ton	0	ton	0		
Concentrated stillage tank agitator	16.00	11.0	ton	0	ton	0		
Process water UV sterilization package	60.00	1.0	EA	0	EA	0	0	0
Diesel generator set; 1000 hp	120.00	1.0	EA	0	EA	0	0	0
Evaporator package; 95′ × 40′ × 80′	320.00	1.0	EA	0	EA	0	0	0

7.40 Wastewater Treatment-Equipment Installation Man Hours

Facility—Ethanol Plant Major Equipment	Actual	Estimate			
	MH	BM	PF	MW	MH
Pumps	347	0	0	0	0
Filters, coolers, tanks, and drums	183	0	0	0	0
Process water UV sterilization package	60	0	0	0	0
Diesel generator set; 1000 hp	120	0	0	0	0
Evaporator package; 95′×40′×80′	320	0	0	0	0
Equipment installation man hours	**1030**	**0**	**0**	**0**	**0**

Utilities

7.41 General Scope of Field Work Required for Utilities

Scope of Work-Field Erection

CW pump #1, #2, and #3; 1750 hp
Chilled water pump #1 and #2; 150 hp
Continuous chiller pump #1 and #2; 125 hp
Fresh water pump; #1 and #2; 50 hp
Demi water pump #1 and #2; 20 hp
Incoming water storage pump 1; 50 hp
Incoming water storage pump 2; 50 hp
CCW expansion tank (shop); 59″ dia×110″; Wt = 3207#
Instrument air buffer; 72″ dia×180″; Wt = 4920#
Compressed air buffer; 102″ dia×180″; Wt = 19,177#
Incoming water treatment system
Multimedia filters; 84″ dia×60″
RO units; 200″L×82″×88″
Water storage tanks; 10″ dia
Antiscalant Portafeed; 48″l×48″ W×48″ H
Air compressor package #1 and #2
Air compressor; 1000 P
Air receiver
Filters
Air compressor package #3
Air compressor; 500 P
Air receiver
Filters
Cooling tower; 94,097 hp; 3 cells (52′×48′); tower 163.5′×54′
Summer chiller #1 and #2; 1333 hp; Wt = 124,031#
Continuous chiller; 2805 hp; Wt = 41,806#
Safety showers

7.42   Utilities Estimate Data

7.42.1   Pumps, Tanks, Filters, Air Compressor, Receiver, Buffer, and Chillers Sheet 1

Description	MH	Unit
Pumps		
Motor horsepower 15–30	1.60	hp
Motor horsepower 76–100	1.00	hp
Motor horsepower less than 15	16.00	EA
Motor horsepower 31–50	1.30	hp
Motor horsepower 101–125	0.90	hp
Motor horsepower 126–200	0.80	hp
Motor horsepower less than 15	16.00	EA
Filters, Coolers, Tanks, and Drums		
CCW expansion tank (shop) and water storage tank	14.50	ton
Multimedia filters, filters	24.00	ton
Air Compressor, Receiver, Air Buffer		
Air Compressor Package #3		
Air compressor; 500 P	60.00	EA
Air receiver	24.00	EA
Air Compressor Package #1 and #2		
Air compressor; 1000 P	120.00	EA
Air receiver	24.00	EA
Instrument air buffer; 72″ dia × 180″; Wt = 4920#	14.50	ton
Compressed air buffer; 102″ dia × 180″; Wt = 19177#	14.50	ton
Chillers		
Summer chiller #1 and #2; 1333 hp; Wt = 124031#	3.77	ton
Continuous chiller; 2805 hp; Wt = 41806#	6.40	ton
Safety showers	16.00	EA
Antiscalant Portafeed; 48′ l × 48′ W × 48′ H	4.00	EA

7.43 Equipment Installation Estimate

7.43.1 Pumps, Tanks, Filters, Air Compressor, Receiver, Buffer, and Chillers Sheet 1

	Historical			Estimate				
	MH	Quantity	Unit	Quantity	Unit	BM	PF	MW
Pumps						0	0	0
CW pump #1, #2, and #3; 1750 hp	0.60	1750.0	hp	0	hp		0	0
Chilled water pump #1 and #2; 150 hp	0.80	300.0	hp	0	hp		0	0
Continuous chiller pump #1 and #2; 125 hp	0.90	250.0	hp	0	hp		0	0
Fresh water pump; #1 and #2; 50 hp	1.30	100.0	hp	0	hp		0	0
Demi water pump #1 and #2; 20 hp	1.60	40.0	hp	0	hp		0	0
Incoming water storage pump 1; 50 hp	1.30	50.0	hp	0	hp		0	0
Incoming water storage pump 2; 50 hp	1.30	50.0	hp	0	hp		0	0
Filters, coolers, tanks, and drums						0	0	0
CCW expansion tank (shop); 59″ dia×110″	14.50	1.6	ton	0	ton	0		
Water storage tanks; 10″ dia	14.50	1.0	ton	0	ton	0		
Multimedia filters; 84″ dia×60″	24.00	1.0	EA	0	EA	0		
Filters	24.00	4.0	EA	0	EA	0		
Air compressor, receiver, air buffer						0	0	0
Air Compressor Package #3								
Air compressor; 500 P	60.00	1.0	EA	0	EA	0		
Air receiver	24.00	2.0	EA	0	EA	0		
Air Compressor Package #1 and #2								
Air compressor; 1000 P	120.00	2.0	EA	0	EA	0		
Air receiver	24.00	2.0	EA	0	EA	0		
Instrument air buffer; 72″ dia×180″	14.50	2.5	ton	0	ton	0		
Compressed air buffer; 102″ dia×180″	14.50	9.6	ton	0	ton	0		
Chillers						0	0	0
Summer chiller #1 and #2; 1333 hp	3.77	124.0	ton	0	ton	0		0
Continuous chiller; 2805 hp	6.40	20.9	ton	0	ton	0		0
Safety showers	16.00	20.0	EA	0	EA	**0**	**0**	**0**
Antiscalant Portafeed; 48′ l×48′ W×48′ H	4.00	2.0	EA	0	EA	**0**	**0**	**0**

7.44   Utilities-Equipment Installation Man Hours

Facility—Ethanol Plant Major Equipment	Actual	Estimate			
	MH	BM	PF	MW	MH
Pumps	1839	0	0	0	0
Filters, coolers, tanks, and drums	158	0	0	0	0
Air compressor, receiver, air buffer	572	0	0	0	0
Chillers	601	0	0	0	0
Safety showers	320	0	0	0	0
Antiscalant Portafeed; 48″ 1×48″ W×48″ H	16	0	0	0	0
Utilities-equipment installation man hours	**3506**	**0**	**0**	**0**	**0**

Package Boiler

7.45   General Scope of Field Work Required for Package Boiler

Scope of Work-Field Erection

Gas boiler package
Boiler 1800 hp; steam 60,000 lbs/h; BTU 60.2 mm BTU
Revolving stoker bed-modular package
Includes:
Watertube section w/pipe
Chain grate stoker system
Combustion chamber
Primary ash collection screw
Oil control campers
Gas/oil back up burner
Over fire system
Ash reinjection blower
Fire doors
Water side inspection ports
Twin screw fuel metering Bins
Fuel supply conveyor
Ladders and platforms
Boiler feed water deaerator package; pumps and deaerator
Stillage boiler

7.46   Equipment Installation Estimate

Estimate Boiler 1800 hp; Steam 60,000 lbs/h; BTU 60.2 mm BTU

		Historical		Estimate				
	MH	Quantity	Unit	Quantity	Unit	BM	PF	MW
Package boiler						0	0	0
Install steam and mud drums	166	1.0	Unit	0	Unit	0		
Generating bank tubes	544	1.0	Unit	0	Unit	0		
Assembly-headers, furnace, and waterwall panels	404	1.0	Unit	0	Unit	0		
Headers and wall panels, fit and weld waterwall panels	665	1.0	Unit	0	Unit	0		
Weld waterwall tubes; 2-1/2″ OD, BW (TIG)	1037	1.0	Unit	0	Unit	0		
Weld primary/ secondary headers and SH elements	227	1.0	Unit	0	Unit	0		
Burner system	63	1.0	Unit	0	Unit	0		
Sootblowers	192	1.0	Unit	0	Unit	0		
Erect and install downcomers and code piping	191	1.0	Unit	0	Unit	0		
Stoker-equipment and piping	335	1.0	Unit	0	Unit	0	0	0
Metering bins	159	1.0	Unit	0	Unit	0	0	0
Boiler drag chain	129	1.0	Unit	0	Unit	0	0	0
Fuel chutes	123	1.0	Unit	0	Unit	0	0	0
Bottom ash drag system	190	1.0	Unit	0	Unit	0	0	0
Structural steel and **boiler casing**						0	0	0
Erect boiler structural steel	690	1.0	Unit	0	Unit	0	0	0
Erect penthouse steel	165	1.0	Unit	0	Unit	0	0	0
Install platform, grating and handrail	921	1.0	Unit	0	Unit	0	0	0
Install stair No. 1 and No. 2	164	1.0	Unit	0	Unit	0	0	0
Casing-generating bank, boiler, and penthouse	177	1.0	Unit	0	Unit	0	0	0

7.47   Package Boiler-Equipment Installation Man Hours

Facility—Ethanol Plant Major Equipment	Actual	Estimate			
	MH	BM	PF	MW	MH
Package boiler	3488	0	0	0	0
Stoker-equipment and piping	335	0	0	0	0
Metering bins	159	0	0	0	0
Boiler drag chain	129	0	0	0	0
Fuel chutes	123	0	0	0	0
Bottom ash drag system	190	0	0	0	0
Structural steel and boiler casing	2117	0	0	0	0
Package boiler-equipment installation man hours	**6542**	**0**	**0**	**0**	**0**

7.48   Ethanol Plant Major Equipment Man Hour Breakdown

Facility—Biomass Plant Major Equipment	Actual	Estimate				
	MH	BM	IW	PF	MW	MH
Biomass handling and bagasse storage installation man hours	2596	0	0		0	0
Pretreatment-equipment installation man hours	3380	0	0		0	0
Viscosity reduction-equipment installation man hours	1360	0		0	0	0
Fermentation-equipment installation man hours	2970	0		0	0	0
Distillation and dehydration-equipment installation man hours	5860	0		0	0	0
Ethanol storage and loading-equipment installation man hours	2580	0		0	0	0
Chemical storage-equipment installation man hours	1382	0		0	0	0
Lignin separation-equipment installation man hours	1080	0	0		0	0
Lignin storage and handling-equipment installation man hours	1314	0		0	0	0
Wastewater treatment-equipment installation man hours	1030	0		0	0	0
Utilities-equipment installation man hours	3506	0		0	0	0
Package boiler-equipment installation man hours	6542	0				0
Ethanol plant project man hours	**33,600**	**0**	**0**	**0**	**0**	**0**

Sample Estimates and Statistical Applications to Construction

8.1 Section Introduction

Sample Estimates

The purpose of this section is to provide the estimator with piping and equipment estimate forms to set up detailed estimates using a desktop computer. Manual estimates made with a pad and pencil are time-consuming. Desktop and network PCs are the hardware for computer-assisted estimating. The piping take-off quantities and the craft unit rates are entered into the estimate forms. The piping estimated craft man hours are calculated using the unit quantity method.

The sample estimate does not include cost and man hours for material, equipment usage, indirect craft and supervision, project staff, warehousing and storage, shop fabrication, overheads, and fee. The direct craft man hour estimate is the basis for the estimator to obtain the project schedule and the man hours and cost for indirect craft and supervision, project staff, construction equipment, material, subcontractors, mobilize and demobilize, site general conditions, overhead, and fee. In addition, the estimator must determine all factors that will affect direct craft labor productivity and overtime impacts. The estimator can use the table of labor factors and values for factoring labor productivity in the Introduction.

Statistical Applications to Construction

Statistics is a body of methods enabling us to draw reasonable conclusions from data.

Statistics is divided into two general types: descriptive statistics and statistical inference.

With descriptive statistics, we summarize data, making calculations, tables, or graphs that can be comprehended easily. Statistical inference, involves drawing conclusions from the data. The purpose of the following sample statistical applications is to make the connection to construction.

To get the most benefits from the statistical applications, the reader should understand exponents, logarithms, and simple algebraic manipulations. Also understanding of regression analysis helps, but not required. Readers whose mathematics is limited can plug the equations into a spreadsheet to get the results they need.

Industrial Piping and Equipment Estimating Manual. http://dx.doi.org/10.1016/B978-0-12-813946-2.00008-3

8.2   Process Piping Estimate Forms

Facility-
  Equipment-
  Estimate Sheet 1 Handle and Install Pipe-Welded Joint

| Description | Size | Pipe Handle | | | PF | Hydro | | PF |
		LF	DIF	MH/ DIF	MH	DIF	MH/ DIF	MH
	0	0	0	0	0	0	0	0
	0	0	0	0	0	0	0	0
	0	0	0	0	0	0	0	0
	0	0	0	0	0	0	0	0
	0	0	0	0	0	0	0	0
	0	0	0	0	0	0	0	0
	0	0	0	0	0	0	0	0
	0	0	0	0	0	0	0	0
	0	0	0	0	0	0	0	0
	0	0	0	0	0	0	0	0
	0	0	0	0	0	0	0	0
	0	0	0	0	0	0	0	0
	0	0	0	0	0	0	0	0
	0	0	0	0	0	0	0	0
	0	0	0	0	0	0	0	0
	0	0	0	0	0	0	0	0
	0	0	0	0	0	0	0	0
	0	0	0	0	0	0	0	0
	0	0	0	0	0	0	0	0
	0	0	0	0	0	0	0	0
Column totals		**0**	**0**		**0**	**0**		**0**

Facility-
  Equipment-
  Estimate Sheet 2 Welding: BW, SW, PWHT Arc-Uphill

| Description | Size | BW | SW | | BW | SW | | PWHT | PF |
		JT	JT	DI	MH/ DI	MH/ JT	Factor	MH/JT	MH
	0	0	0	0	0	0	0	0	0
	0	0	0	0	0	0	0	0	0
	0	0	0	0	0	0	0	0	0
	0	0	0	0	0	0	0	0	0
	0	0	0	0	0	0	0	0	0
	0	0	0	0	0	0	0	0	0
	0	0	0	0	0	0	0	0	0

Description	Size	BW JT	SW JT	DI	BW MH/DI	SW MH/JT	Factor	PWHT MH/JT	PF MH
	0	0	0	0	0	0	0	0	0
	0	0	0	0	0	0	0	0	0
	0	0	0	0	0	0	0	0	0
	0	0	0	0	0	0	0	0	0
	0	0	0	0	0	0	0	0	0
	0	0	0	0	0	0	0	0	0
	0	0	0	0	0	0	0	0	0
	0	0	0	0	0	0	0	0	0
	0	0	0	0	0	0	0	0	0
	0	0	0	0	0	0	0	0	0
	0	0	0	0	0	0	0	0	0
	0	0	0	0	0	0	0	0	0
	0	0	0	0	0	0	0	0	0
Column totals		0	0	0					0

Facility-

Equipment-

Estimate Sheet 3 Bolt Up of Flanged Joint by Weight Class

Description	Size	150#/300# Bolt Up	600#/900# Bolt Up	1500#/2500# Bolt Up	DI	MH/DI	MH/DI	MH/DI	PF MH
	0	0	0	0	0	0	0	0	0
	0	0	0	0	0	0	0	0	0
	0	0	0	0	0	0	0	0	0
	0	0	0	0	0	0	0	0	0
	0	0	0	0	0	0	0	0	0
	0	0	0	0	0	0	0	0	0
	0	0	0	0	0	0	0	0	0
	0	0	0	0	0	0	0	0	0
	0	0	0	0	0	0	0	0	0
	0	0	0	0	0	0	0	0	0
	0	0	0	0	0	0	0	0	0
	0	0	0	0	0	0	0	0	0
	0	0	0	0	0	0	0	0	0
	0	0	0	0	0	0	0	0	0
	0	0	0	0	0	0	0	0	0
	0	0	0	0	0	0	0	0	0
	0	0	0	0	0	0	0	0	0
	0	0	0	0	0	0	0	0	0
	0	0	0	0	0	0	0	0	0
	0	0	0	0	0	0	0	0	0
Column totals					0				0

Facility-
Equipment-
Estimate Sheet 4: Handle Valves by Weight Class

Description	Size	150#/ 300# Valve	600#/ 900# Valve	1500#/ 2500# Valve	DI	MH/ DI	MH/ DI	MH/ DI	PF MH
		0	0	0	0	0	0	0	0
		0	0	0	0	0	0	0	0
		0	0	0	0	0	0	0	0
		0	0	0	0	0	0	0	0
		0	0	0	0	0	0	0	0
		0	0	0	0	0	0	0	0
		0	0	0	0	0	0	0	0
		0	0	0	0	0	0	0	0
		0	0	0	0	0	0	0	0
		0	0	0	0	0	0	0	0
		0	0	0	0	0	0	0	0
		0	0	0	0	0	0	0	0
		0	0	0	0	0	0	0	0
		0	0	0	0	0	0	0	0
		0	0	0	0	0	0	0	0
		0	0	0	0	0	0	0	0
		0	0	0	0	0	0	0	0
		0	0	0	0	0	0	0	0
Column totals		0	0	0	0	0	0	0	0

Facility-
Equipment-
Estimate Sheet 5: Pipe Supports

Description	Material	Size	Sch/Thk	Pipe Support	DI	MH/DI	PF MH
				0	0	0	0
				0	0	0	0
				0	0	0	0
				0	0	0	0
				0	0	0	0
				0	0	0	0
				0	0	0	0
				0	0	0	0

Description	Material	Size	Sch/Thk	Pipe Support	DI	MH/DI	PF MH
				0	0	0	0
				0	0	0	0
				0	0	0	0
				0	0	0	0
				0	0	0	0
				0	0	0	0
				0	0	0	0
				0	0	0	0
				0	0	0	0
				0	0	0	0
				0	0	0	0
Column totals				**0**	**0**		**0**

Facility-
Equipment-
Estimate Sheet 6: Instrument

Description	Material	Size	Sch/Thk	Instrument	MH/EA	PF MH
				0	0.00	0
				0	0.00	0
				0	0.00	0
				0	0.00	0
				0	0.00	0
				0	0.00	0
				0	0.00	0
				0	0.00	0
				0	0.00	0
				0	0.00	0
				0	0.00	0
				0	0.00	0
				0	0.00	0
				0	0.00	0
				0	0.00	0
				0	0.00	0
				0	0.00	0
				0	0.00	0
				0	0.00	0
Column totals				**0**		**0**

Facility-
Equipment-
Estimate Sheet 7: Summary

	PF	MH/LF
Description	**MH**	
Estimate sheet 1: handle and install pipe-welded joint	0	
Estimate sheet 2: welding: BW, SW, PWHT arc-uphill	0	
Estimate sheet 3: bolt up of flanged joint by weight class	0	
Estimate sheet 4: handle valves by weight class	0	
Estimate sheet 5: pipe supports	0	
Estimate sheet 6: instrument	**0**	
Column totals	**0**	0.00

8.3   Sample Process Piping and Equipment Estimates

8.3.1   Field Erect HRSG HP Piping and Supports

Facility-Combined Cycle Power Plant

Equipment—Triple Pressure w/Reheat for F-Class GT-Three Wide

Estimate Sheet 1: Handle and Install Pipe-Welded Joint

		Pipe Handle			PF	Hydro		PF	Total
Description	**Size**	**LF**	**DIF**	**MH/ DIF**	**MH**	**DIF**	**MH/ DIF**	**MH**	**MH**
HP Piping and Supports									
Sch 140/0.906″, SA-106 B, HP-03 Econ #1 to HP Econ #2	8	25	200	0.14	28	200	0.134	27	55
Sch 160/0.906″, SA-106 B, HP-03 Econ #1 to HP Econ #2	6	42	252	0.14	35	252	0.101	25	61
Sch 140/0.812″, SA-106 B, HP-04 Econ #2 to HP steam drum	8	96	768	0.14	108	768	0.134	103	211
Sch 160/0.906″, SA-106 C, HP-04 Econ #2 to HP steam drum	6	1	6	0.14	1	6	0.101	1	1

| Description | Size | Pipe Handle | | | PF | Hydro | | PF | Total |
		LF	DIF	MH/ DIF	MH	DIF	MH/ DIF	MH	MH
Sch 160/0.906″, SA-106 B, HP-09 steam drum to HP SH #1	6	24	144	0.14	20	144	0.101	15	35
Sch 160/0.906″, SA-106 B, HP-10 steam drum to HP SH #1	6	23	138	0.14	19	138	0.101	14	33
Sch 160/0.906″, SA-106 B, HP-11 steam drum to HP SH #1	6	23	138	0.14	19	138	0.101	14	33
Sch 160/0.906″, SA-106 B, HP-12 steam drum to HP SH #1	6	23	138	0.14	19	138	0.101	14	33
Sch 160/0.906″, SA-106 B, HP-13 steam drum to HP SH #1	6	23	138	0.14	19	138	0.101	14	33
Sch 160/0.906″, SA-106 B, HP-14 steam drum to HP SH #1	6	23	138	0.14	19	138	0.101	14	33
1.25″ WT, SA-335 P91, HP-15 SHTR #1 to HP SHTR #2	10	68	680	0.25	170	680	0.240	163	333
1.031″ WT, SA-335 P91, HP-15 SHTR #1 to HP SHTR #2	8	3	24	0.20	5	24	0.192	5	9
SA-335 P91, HP-15 SHTR #1 to HP SHTR #2	3	1	3	0.20	1	3		0	1

Continued

Description	Size	Pipe Handle LF	DIF	MH/ DIF	PF MH	Hydro DIF	MH/ DIF	PF MH	Total MH
1.25" WT, SA-335 P91, HP-16 SHTR #1 to HP SHTR #2	10	68	680	0.25	170	680	0.240	163	333
1.032" WT, SA-335 P91, HP-16 SHTR #1 to HP SHTR #2	8	3	24	0.20	5	24	0.192	5	9
SA-335 P91, HP-16 SHTR #1 to HP SHTR #2	3	1	3	0.20	1	3		0	1
1.25" WT, SA-335 P91, HP-17 SHTR #1 to HP SHTR #2	10	68	680	0.25	170	680	0.240	163	333
1.031" WT, SA-335 P91, HP-17 SHTR #1 to HP SHTR #2	8	3	24	0.20	5	24	0.192	5	9
SA-335 P91, HP-17 SHTR #1 to HP SHTR #2	3	1	3	0.20	1	3		0	1
Column totals		519	4181		815	4181		744	1558

Facility-Combined Cycle Power Plant

Equipment—Triple Pressure w/Reheat for F-Class GT-Three Wide

Estimate Sheet 2: Welding: BW, SW, PWHT Arc-Uphill

Description	Size	BW JT	SW JT	DI	BW MH/ DI	SW MH/ JT	Factor	PWHT MH/JT	PF MH
HP Piping and Supports									
Sch 140/0.906", SA-106 B, HP-03 Econ #1 to HP Econ #2	8			0	1.20	1.4	1.0	0	0
Sch 160/0.906", SA-106 B, HP-03 Econ #1 to HP Econ #2	6	9	9	54	1.20	1.4	1.0	0.45	102

Description	Size	BW JT	SW JT	DI	BW MH/ DI	SW MH/ JT	Factor	PWHT MH/JT	PF MH
Sch 140/0.812″, SA-106 B, HP-04 Econ #2 to HP steam drum	8	4	6	32	1.20	1.4	1.0	0.45	61
Sch 160/0.906″, SA-106 C, HP-04 Econ #2 to HP steam drum	6	3		18	1.20	1.4	1.0	0.45	30
Sch 160/0.906″, SA-106 B, HP-09 steam drum to HP SH #1	6	4	4	24	1.20	1.4	1.0	0.45	45
Sch 160/0.906″, SA-106 B, HP-10 steam drum to HP SH #1	6	3	4	18	1.20	1.4	1.0	0.45	35
Sch 160/0.906″, SA-106 B, HP-11 steam drum to HP SH #1	6	4	4	24	1.20	1.4	1.0	0.45	45
Sch 160/0.906″, SA-106 B, HP-12 steam drum to HP SH #1	6	3	4	18	1.20	1.4	1.0	0.45	35
Sch 160/0.906″, SA-106 B, HP-13 steam drum to HP SH #1	6	4	4	24	1.20	1.4	1.0	0.45	45
Sch 160/0.906″, SA-106 B, HP-14 steam drum to HP SH #1	6	3	4	18	1.20	1.4	1.0	0.45	35
1.25″ WT, SA-335 P91, HP-15 SHTR #1 to HP SHTR #2	10	4	5	40	2.2	1.4	2.8	0.45	279

Continued

		BW	SW		BW	SW		PWHT	PF
Description	Size	JT	JT	DI	MH/ DI	MH/ JT	Factor	MH/JT	MH
1.031″ WT, SA-335 P91, HP-15 SHTR #1 to HP SHTR #2	8	2		16	1.45	1.4	2.5	0.45	65
SA-335 P91, HP-15 SHTR #1 to HP SHTR #2	3			0	0	1.4	1.5	0	0
1.25″ WT, SA-335 P91, HP-16 SHTR #1 to HP SHTR #2	10	4	5	40	2.2	1.4	2.8	0.45	279
1.032″ WT, SA-335 P91, HP-16 SHTR #1 to HP SHTR #2	8	2		16	1.45	1.4	2.5	0.45	65
SA-335 P91, HP-16 SHTR #1 to HP SHTR #2	3	0	0	0	1.20	1.4	1.5		0
1.25″ WT, SA-335 P91, HP-17 SHTR #1 to HP SHTR #2	10	4	5	40	2.2	1.4	2.8	0.45	279
1.031″ WT, SA-335 P91, HP-17 SHTR #1 to HP SHTR #2	8	2		16	1.45	1.4	2.5	0.45	65
SA-335 P91, HP-17 SHTR #1 to HP SHTR #2	3			0			1.5		0
Column totals		55	54	398					1467

Facility-Combined Cycle Power Plant

Equipment—Triple Pressure w/Reheat for F-Class GT-Three Wide

Estimate Sheet 3: Bolt Up of Flanged Joint by Weight Class

		150#/300#	600#/900#	1500#/2500#					PF
Description	**Size**	**Bolt Up**	**Bolt Up**	**Bolt Up**	**DI**	**MH/DI**	**MH/DI**	**MH/DI**	**MH**
HP Piping and Supports									
HP-03 Econ #1 to HP Econ #2	8	0	0	0	0	0.40	0.50	0.65	0
HP-03 Econ #1 to HP Econ #2	6	0	0	0	0	0.40	0.50	0.65	0
HP-04 Econ #2 to HP steam drum	8	0	0	0	0	0.40	0.50	0.65	0
HP-04 Econ #2 to HP steam drum	6	0	0	0	0	0.40	0.50	0.65	0
HP-09 steam drum to HP SH #1	6	0	0	0	0	0.40	0.50	0.65	0
HP-10 steam drum to HP SH #1	6	0	0	0	0	0.40	0.50	0.65	0
HP-11 steam drum to HP SH #1	6	0	0	0	0	0.40	0.50	0.65	0
HP-12 steam drum to HP SH #1	6	0	0	0	0	0.40	0.50	0.65	0
HP-13 steam drum to HP SH #1	6	0	0	0	0	0.40	0.50	0.65	0
HP-14 steam drum to HP SH #1	6	0	0	0	0	0.40	0.50	0.65	0
HP-15 SHTR #1 to HP SHTR #2	10	0	0	0	0	0.40	0.50	0.65	0
HP-15 SHTR #1 to HP SHTR #2	8	0	0	0	0	0.40	0.50	0.65	0
HP-15 SHTR #1 to HP SHTR #2	3	0	0	0	0	0.40	0.50	0.65	0
HP-16 SHTR #1 to HP SHTR #2	10	0	0	0	0	0.40	0.50	0.65	0
HP-16 SHTR #1 to HP SHTR #2	8	0	0	0	0	0.40	0.50	0.65	0
HP-16 SHTR #1 to HP SHTR #2	3	0	0	0	0	0.40	0.50	0.65	0
HP-17 SHTR #1 to HP SHTR #2	10	0	0	0	0	0.40	0.50	0.65	0
HP-17 SHTR #1 to HP SHTR #2	8	0	0	0	0	0.40	0.50	0.65	0
HP-17 SHTR #1 to HP SHTR #2	3	0	0	0	0	0.40	0.50	0.65	0
Column totals					**0**				**0**

Facility-Combined Cycle Power Plant
Equipment—Triple Pressure w/Reheat for F-Class GT-Three Wide
Estimate Sheet 4: Handle Valves by Weight Class

HP Piping and Supports	Size	Valve	Valve	Valve	DI	MH/DI	MH/DI	MH/DI	MH
Description		150#/300#	600#/900#	1500#/2500#					PF
HP-03 Econ #1 to HP Econ #2	8	0	0	0	0	0.45	0.90	1.8	0
HP-03 Econ #1 to HP Econ #2	6	0	0	0	0	0.45	0.90	1.8	0
HP-04 Econ #2 to HP steam drum	8	0	0	0	0	0.45	0.90	1.8	0
HP-04 Econ #2 to HP steam drum	8	0	0	0	0	0.45	0.90	1.8	0
HP-09 steam drum to HP SH #1	6	0	0	0	0	0.45	0.90	1.8	0
HP-10 steam drum to HP SH #1	6	0	0	0	0	0.45	0.90	1.8	0
HP-11 steam drum to HP SH #1	6	0	0	0	0	0.45	0.90	1.8	0
HP-12 steam drum to HP SH #1	6	0	0	0	0	0.45	0.90	1.8	0
HP-13 steam drum to HP SH #1	6	0	0	0	0	0.45	0.90	1.8	0
HP-14 steam drum to HP SH #1	6	0	0	0	0	0.45	0.90	1.8	0
HP-15 SHTR #1 to HP SHTR #2	6	0	0	0	0	0.45	0.90	1.8	0
HP-15 SHTR #1 to HP SHTR #2	10	0	0	0	0	0.45	0.90	1.8	0
HP-15 SHTR #1 to HP SHTR #2	8	0	0	0	0	0.45	0.90	1.8	0
HP-16 SHTR #1 to HP SHTR #2	3	0	0	0	0	0.45	0.90	1.8	0
HP-16 SHTR #1 to HP SHTR #2	10	0	0	0	0	0.45	0.90	1.8	0
HP-16 SHTR #1 to HP SHTR #2	8	0	0	0	0	0.45	0.90	1.8	0
HP-17 SHTR #1 to HP SHTR #2	3	0	0	0	0	0.45	0.90	1.8	0
HP-17 SHTR #1 to HP SHTR #2	10	0	0	0	0	0.45	0.90	1.8	0
HP-17 SHTR #1 to HP SHTR #2	8	0	0	0	0	0.45	0.90	1.8	0
HP-17 SHTR #1 to HP SHTR #2	3	0	0	0	0	0.45	0.90	1.8	0
Column totals					0				0

Facility-Combined Cycle Power Plant
Equipment—Triple Pressure w/Reheat for F-Class GT-Three Wide
Estimate Sheet 5: Pipe Supports

Description	Material	Size	Sch/Thk	Pipe		MH/DI	PF
				Support	DI		MH
HP Piping and Supports							
HP-03 Econ #1 to HP Econ #2	SA-106-B	8	140/.906		0	1.00	0
HP-03 Econ #1 to HP Econ #2	SA-106-B	6	160/.906		0	1.00	0
HP-04 Econ #2 to HP steam drum	SA-106-B	8	140/.812	2	16	1.00	16
HP-04 Econ #2 to HP steam drum	SA-106-C	6	160/.906		0	1.00	0
HP-09 steam drum to HP SH #1	SA-106-B	6	160/.906	1	6	1.00	6
HP-10 steam drum to HP SH #1	SA-106-B	6	160/.906	1	6	1.00	6
HP-11 steam drum to HP SH #1	SA-106-B	6	160/.906	1	6	1.00	6
HP-12 steam drum to HP SH #1	SA-106-B	6	160/.906	1	6	1.00	6
HP-13 steam drum to HP SH #1	SA-106-B	6	160/.906	1	6	1.00	6
HP-14 steam drum to HP SH #1	SA-106-B	6	160/.906	1	6	1.00	6
HP-15 SHTR #1 to HP SHTR #2	SA-335-P91	10	1.25	2	20	1.00	20
HP-15 SHTR #1 to HP SHTR #2	SA-335-P91	8	1.031		0	1.00	0
HP-15 SHTR #1 to HP SHTR #2	SA-335-P91	3			0	1.00	0
HP-16 SHTR #1 to HP SHTR #2	SA-335-P91	10	1.25		0	1.00	0
HP-16 SHTR #1 to HP SHTR #2	SA-335-P91	8	1.031		0	1.00	0
HP-16 SHTR #1 to HP SHTR #2	SA-335-P91	3			0	1.00	0
HP-17 SHTR #1 to HP SHTR #2	SA-335-P91	10	1.25	2	20	1.00	20
HP-17 SHTR #1 to HP SHTR #2	SA-335-P91	8	1.031		0	1.00	0
HP-17 SHTR #1 to HP SHTR #2	SA-335-P91	3			0	1.00	0
Column totals				12	92		92

Facility-Combined Cycle Power Plant Equipment—Triple Pressure w/Reheat for F-Class GT-Three Wide

Estimate Sheet 6: Instrument

Description	Material		Size	Sch/Thk	Instrument	MH/EA	PF	MH
HP Piping and Supports								
HP-03 Econ #1 to HP Econ #2	SA-106-B	8	140/.906		1.20	0		
HP-03 Econ #1 to HP Econ #2	SA-106-B	6	160/.906	9	1.20	10.8		
HP-04 Econ #2 to HP steam drum	SA-106-B	8	140/.812	6	1.20	7.2		
HP-04 Econ #2 to HP steam drum	SA-106-C	6	160/.906		1.20	0		
HP-09 steam drum to HP SH #1	SA-106-B	6	160/.906	4	1.20	4.8		
HP-10 steam drum to HP SH #1	SA-106-B	6	160/.906	4	1.20	4.8		
HP-11 steam drum to HP SH #1	SA-106-B	6	160/.906	4	1.20	4.8		
HP-12 steam drum to HP SH #1	SA-106-B	6	160/.906	4	1.20	4.8		
HP-13 steam drum to HP SH #1	SA-106-B	6	160/.906	4	1.20	4.8		
HP-14 steam drum to HP SH #1	SA-106-B	6	160/.906	4	1.20	4.8		
HP-15 SHTR #1 to HP SHTR #2	SA-335-P91	10	1.25	5	1.20	6		
HP-15 SHTR #1 to HP SHTR #2	SA-335-P91	8	1.031		1.20	0		
HP-15 SHTR #1 to HP SHTR #2	SA-335-P91	3			1.20	0		
HP-16 SHTR #1 to HP SHTR #2	SA-335-P91	10	1.25	5	1.20	6		
HP-16 SHTR #1 to HP SHTR #2	SA-335-P91	8	1.031		1.20	0		
HP-16 SHTR #1 to HP SHTR #2	SA-335-P91	3			1.20	0		
HP-17 SHTR #1 to HP SHTR #2	SA-335-P91	10	1.25	5	1.20	6		
HP-17 SHTR #1 to HP SHTR #2	SA-335-P91	8	1.031		1.20	0		
HP-17 SHTR #1 to HP SHTR #2	SA-335-P91	3			1.20	0		
Column totals				54		64.8		

Facility-Combined Cycle Power Plant
Equipment—Triple Pressure w/Reheat for F-Class GT-Three Wide
Estimate Sheet 7: Summary HP Piping and Supports

Description	PF	MH/LF
	MH	
Estimate sheet 1: handle and install pipe-welded joint	1558	
Estimate sheet 2: welding: BW, SW, PWHT arc-uphill	1467	
Estimate sheet 3: bolt up of flanged joint by weight class	0	
Estimate sheet 4: handle valves by weight class	0	
Estimate sheet 5: pipe supports	92	
Estimate sheet 6: instrument	**65**	
Column totals	**3182**	6.13

8.4 Equipment Estimate Form

Estimate-

　Facility-

Description	Historical			Estimate					
	MH	Quantity	Unit	Quantity	Unit	BM	IW	PF	MW
	0.00	0.0		0.0		0	0	0	0
	0.00	0.0		0.0		0	0	0	0
	0.00	0.0		0.0		0	0	0	0
	0.00	0.0		0.0		0	0	0	0
	0.00	0.0		0.0		0	0	0	0
	0.00	0.0		0.0		0	0	0	0
	0.00	0.0		0.0		0	0	0	0
	0.00	0.0		0.0		0	0	0	0
	0.00	0.0		0.0		0	0	0	0
	0.00	0.0		0.0		0	0	0	0
	0.00	0.0		0.0		0	0	0	0

　Facility-

Equipment-Installation Man Hours

Description	Actual	Estimated
	MH	BM
	0	0
	0	0
	0	0

Continued

Description	Actual MH	Estimated BM
	0	0
	0	0
	0	0
	0	0
	0	0
	0	0
Equipment installation man hours	**0**	**0**

8.4.1   Field Erect HRSG Double Pressure Single Wide

HRSG Double Pressure Single Wide Estimate
   Modules Box 1, 2, 3, 4, Spool Duct and **Casing Sheet 1**
   Facility-Combined Cycle Power Plant

Description	MH	Historical Quantity	Unit	Estimate Quantity	Unit	BM
Modules Box 1, 2, 3, 4, Spool Duct and Casing Sheet 1						2506
Base plate (BP)	32.00	5.0	EA	5.0	EA	160
Box #1	1.50	148.0	ton	148.0	ton	222
Box #2	1.50	134.0	ton	134.0	ton	201
Box #3	1.50	124.0	ton	124.0	ton	186
Box #4	1.50	130.0	ton	130.0	ton	195
Box 1, 2, 3 and 4 weld to base plate	1.50	16.0	EA	16.0	EA	24
Remove shipping steel box 1, 2, 3, and 4	2.00	16.0	EA	16.0	EA	32
Spool Duct/Casing/Field Joint/Box 2, 3, and 4						
Between Box 2 and Box 3						
Duct bottom transverse beam	5.50	2.0	EA	2.0	EA	11
Duct top transverse beam	5.50	2.0	EA	2.0	EA	11
Moment weld	5.50	4.0	EA	4.0	EA	22
Install floor and roof panels	5.50	4.0	EA	4.0	EA	22
Temporary support	16.00	1.0	EA	1.0	EA	16
Install bottom and top cover Plate	5.50	2.0	EA	2.0	EA	11
Bolt bottom and top cover plate to transverse beam	0.15	76.0	EA	76.0	EA	11.4
Weld top and bottom cover plates	0.35	88.0	LF	88.0	LF	30.9
Install side panels						
Bolt side panels to columns and beams	0.15	536.0	EA	536.0	EA	80.4

Description	Historical			Estimate		
	MH	Quantity	Unit	Quantity	Unit	BM
Seal weld bolts	0.35	536.0	LF	536.0	LF	188.6
Seal weld panels	0.35	340.0	LF	340.0	LF	119.6
Between Box 3 and Box 4						
Duct bottom transverse beam	5.50	2.0	EA	2.0	EA	11
Duct top transverse beam	5.50	2.0	EA	2.0	EA	11
Moment weld	5.50	4.0	EA	4.0	EA	22
Install floor and roof panels	5.50	4.0	EA	4.0	EA	22
Temporary support	16.00	1.0	EA	1.0	EA	16
Install bottom and top cover plate	5.50	2.0	EA	2.0	EA	11
Bolt bottom and top cover plate to transverse beam	0.15	48.0	EA	48.0	EA	7.2
Seal weld top and bottom cover plates	0.35	88.0	LF	88.0	LF	30.9
Install side panels						
Bolt side panels to columns and beams	0.15	536.0	EA	536.0	EA	80.4
Seal weld bolts	0.35	536.0	LF	536.0	LF	188.6
Seal weld panels	0.35	340.0	LF	340.0	LF	119.6
SCR Duct/Field Joint						
Install floor and roof	5.50	2.0	EA	2.0	EA	11
Bolt floor and roof to columns	0.15	108.0	EA	108.0	EA	16.2
Seal weld floor and roof panels	0.35	44.0	LF	44.0	LF	15.4
Install side panels	5.50	3.0	EA	3.0	EA	16.5
Bolt side panels to columns	0.15	496.0	EA	496.0	EA	74.4
Seal weld bolts	0.35	496.0	LF	496.0	LF	173.6
Seal weld panels	0.35	328.0	LF	328.0	LF	114.8
SCR weld to base plate	10.00	2.0	EA	2.0	EA	20

Inlet Duct, Burner, and Distribution Grid Sheet 2
Facility-Combined Cycle Power Plant

Description	Historical			Estimate		
	MH	Quantity	Unit	Quantity	Unit	BM
Inlet Duct, Burner, and Distribution Grid Sheet 2						595
Duct column A-B	24.00	1.0	EA	1.0	EA	24
Field joint at B bolt and seal weld	0.35	55.0	LF	55.0	LF	19.2
Install insulation and liners	65.00	1.0	EA	1.0	EA	65
Duct column B-C	24.00	1.0	EA	1.0	EA	24
Field joint at C bolt and seal weld	0.35	80.0	LF	80.0	LF	28
Install insulation and liners	65.00	1.0	EA	1.0	EA	65
Duct column C-D	24.00	1.0	EA	1.0	EA	24

Continued

Description	Historical			Estimate		
	MH	Quantity	Unit	Quantity	Unit	BM
Field joint at D bolt and seal weld	0.35	144.0	LF	144.0	LF	50.4
Install insulation and liners	65.00	1.0	EA	1.0	EA	65
Inlet duct weld to base plate	5.50	6.0	EA	6.0	EA	33
Duct Burner Assembly						
Main Gas, IA, Igniter Gas, and Cooling Air Headers						
Main gas header	**20.00**	1.0	EA	1.0	EA	20
Cooling air header	**18.00**	1.0	EA	1.0	EA	18
Ignition gas header	**12.00**	1.0	EA	1.0	EA	12
Instrument air header	**20.00**	1.0	EA	1.0	EA	20
Bolt headers to elements	**24.00**	1.0	EA	1.0	EA	24
Distribution Grid						
Grid panels bolt/nut/washer— tack weld	4.40	9.0	EA	9.0	EA	39.6
Install Access Door						
Splice—grid panel to panel 3/32 fillet weld	0.35	45.0	EA	45.0	EA	15.7
Channel	2.20	4.0	EA	4.0	EA	8.8
Support pipe	2.20	4.0	EA	4.0	EA	8.8
Land grid—strut bolt	2.20	8.0	EA	8.0	EA	17.6
Weld grid panel end overlap at support pipe	2.20	6.0	EA	6.0	EA	13.2

Facility-Combined Cycle Power Plant

Estimate—Liner Panels

Description	Historical			Estimate		
	MH	Quantity	Unit	Quantity	Unit	BM
Install Liner Panels Sheet 3						**1040**
Liner panels/bolt	3.39	146.0	EA	146.0	EA	494.6
Liner panels/bolt	3.39	100.0	EA	100.0	EA	338.8
Liner panels/bolt	3.39	61.0	EA	61.0	EA	206.6
Estimate—Vessels	**Historical**			**Estimate**		
Vessels—Drums/Deaerator/ Blowdown Tank Sheet 4						**472**
72″ ID HP steam drum	80.00	1.0	EA	1.0	EA	80
Drum saddle support	5.50	1.0	EA	1.0	EA	5.5
60″ ID LP steam drum	80.00	1.0	EA	1.0	EA	80
Drum saddle support Connection to drum	5.50	1.0	EA	1.0	EA	5.5

Estimate—Vessels	Historical			Estimate		
Deaerator 10'-6" OD × 27'-7" 38,903 lb	60.00	1.0	EA	1.0	EA	60
Fixed supports	8.80	2.0	EA	2.0	EA	17.6
Tank 9'-0" × 5'-0"	12.00	1.0	EA	1.0	EA	12
Blowdown tank 11'-7/8" × 5'-0" ID	12.00	1.0	EA	1.0	EA	12
Make up anchor bolts—grout base plates	8.80	4.0	EA	4.0	EA	35.2
Cems	60.00	1.0	EA	1.0	EA	60
BMS	80.00	1.0	EA	1.0	EA	80
Heat exchanger	24.00	1.0	EA	1.0	EA	24

Estimate—Skids	Historical			Estimate		
Skids Sheet 5						**160**
Fuel skid	80.00	1.0	EA	1.0	EA	80
Blower skid	80.00	1.0	EA	1.0	EA	80

Estimate—Crossover Piping	Historical			Estimate		
Crossover Piping Sheet 6						**54**
Top						
5" s120 CS Pipe x'-x"	1.54	7.0	EA	7.0	EA	10.7
5" s120 CS BW	6.25	4.0	EA	4.0	EA	25
Bottom						
5" s120 CS Pipe 3'-3"	1.54	4.0	EA	4.0	EA	6.16
5" s120 CS BW	6.25	2.0	EA	2.0	EA	12.5

Estimate—Pipe Rack, DA Storage Tank Silencer

Facility-Combined Cycle Power Plant

		Historical		Estimate		
Description	MH	Quantity	Unit	Quantity	Unit	BM
Pipe Rack/DA Storage Tank Silencer Sheet 7						**428**
Pipe Rack	41.30	**7.8**	ton	7.8	ton	320.1
DA Storage Tank Silencer						
W4 × 13 31'	24.00	0.8	ton	0.8	ton	19.3
W6 × 15 4'	24.00	0.2	ton	0.2	ton	4.32
W4 × 13 3'	24.00	0.2	ton	0.2	ton	5.616
Bolt up 9	0.35	180.0	EA	180.0	EA	63
Base plates 3/16" fillet weld/ grout	4.00	4.0	EA	4.0	EA	16

Continued

Estimate—Pumps	Historical			Estimate		
HP and LP Boiler Feed **Pumps Sheet 8**						**240**
HP feed pump 700 hp set/ couple/grout	80.00	3.0	EA	3.0	EA	240

Estimate—Platforms	Historical			Estimate		
Platforms Sheet 9						**1423**
Main deck	48.00	**6.4**	ton	6.4	ton	306.1
Mid level	48.00	**1.9**	ton	1.9	ton	91.5
Lower level	48.00	**2.9**	ton	2.9	ton	139.6
HP and LP drums	48.00	**1.0**	ton	1.0	ton	48
Stack	48.00	**1.1**	ton	1.1	ton	52.3
Deaerator and storage tank	48.00	**3.2**	ton	3.2	ton	151.6
Stair tower	48.00	**13.2**	ton	13.2	ton	633.6

Estimate—Outlet Duct, Stack	Historical			Estimate		
Outlet Duct t/Expansion **Joint/Stack Sheet 10**						**537**
Stack 10'-1" diameter × 118'-10"						
Section 1 39'-9-7/8"	22.00	1.0	EA	1.0	EA	22
Section 2 28'-11-3/8"	22.00	1.0	EA	1.0	EA	22
Section 1 48'-5-3/4"	22.00	3.0	EA	3.0	EA	66
Field weld horizontal	12.00	1.0	EA	1.0	EA	12
Damper						
Inlet flange	11.00	2.0	EA	2.0	EA	22
Plate 4'-8-1/2" × 52'-9-1/4"	11.00	2.0	EA	2.0	EA	22
Plate 4'-8-1/2" × 10'-6"						
Bolt up	0.15	224.0	EA	224.0	EA	33.6
Seal weld 3/16" fillet weld 52'-1"	0.35	125.0	LF	125.0	LF	43.7
Bolt up at foundation AB	0.15	28.0	EA	28.0	EA	4.2
Expansion Joint						
Flex element	22.00	1.0	EA	1.0	EA	22
Retaining bars (2 sets)						
Fasteners	0.03	1004.0	EA	1004.0	EA	33.1
Drop wrap gasket	0.02	460.0	EA	460.0	EA	10.1
3/16" fillet weld	0.35	125.0	LF	125.0	LF	43.7
Hydro	180.00	1.0	test	1.0	test	180

Facility-Combined Cycle Power Plant

HRSG Double Pressure; Single Wide-Installation Man Hours

	Actual	Estimated
Description	**MH**	**BM**
Modules box 1, 2, 3, 4, spool duct and casing sheet 1	2506	2506
Inlet duct, burner, and distribution grid sheet 2	595	595
Install liner panels sheet 3	1040	1040
Vessels—drums/deaerator/blowdown tank sheet 4	472	472
Skids sheet 5	160	160
Crossover piping sheet 6	54	54
Pipe rack/DA storage tank silencer sheet 7	428	428
HP and LP boiler feed pumps sheet 8	240	240
Platforms sheet 9	1423	1423
Outlet duct t/expansion joint/stack sheet 10	537	537
Equipment installation man hours	**7456**	**7456**

8.5   Statistical Applications to Construction

8.5.1   Straight Line Graph; Handle and Install Large Bore Standard Pipe

The graph is a pictorial illustration of the relationship between variables.

$$y = a + bx ; \quad Y = a + (y - y1) / (x - x1)(x)$$

where y = dependent variable; a = intercept value along the y axis at x = 0; b = slope, or the length of the rise divided by the length of the run; $b = (y - y1)/(x - x1)$; x = independent or control variable.

Table 8.1   **Handle and Install Pipe, CS, Welded Joint**

Pipe Size	MH/LF
2.5	0.18
3	0.21
4	0.28
6	0.42
8	0.56
10	0.70
12	0.84
14	0.98
16	1.12
18	1.26
20	1.40
24	1.68

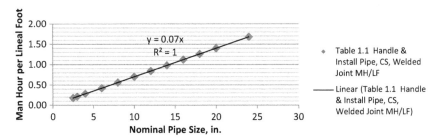

Figure 8.1 Illustration of MH/LF to handle and install large bore standard pipe.

The coefficient of determination, R^2 is exactly +1 and indicates a positive fit.

All data points lie exactly on the straight line. The relationship between X and Y variables is such that as X increases, Y also increases.

8.5.2   Correlation Coefficient and Coefficient of Determination

Correlation Coefficient

1. It measures the strength and direction of a linear relationship between two variables.
2. Mathematical formula for computing r or R is:

$$r = n\left(\sum XY\right) - \left(\sum X\right)\left(\sum Y\right) \Big/ \left(n\left(\sum x^2 - \left(\sum x\right)^2\right)\right)$$
$$\tfrac{1}{2}\left(n\left(\sum y^2 - \left(\sum Y\right)^2\right)\right)\tfrac{1}{2}$$

3. The value of r is such that $(-1 <= r <= +1)$. The + and − signs are used for positive and negative linear correlations, respectively.
4. Positive correlation: If x and y have a strong positive linear correlation, r is close to +1. An r valve exactly +1 indicates a perfect positive fit. Positive values indicate a relationship between X and Y variables such that as values for X increase, values for Y also increase.
5. A perfect correlation of + or − 1 occurs only when all the data points all lie exactly on a straight line. If r = +1, the slope line is positive. If r = −1, the slope if this line is negative.
6. A correlation greater than 0.8 is generally described as strong, whereas a correlation less than 0.5 is generally described as weak.

Coefficient of Determination, r^2 or R^2

1. It is useful because it gives the proportion of variance of one variable that is predictable from the other variable. It is a measure that allows us to determine how certain one can be of making predictions from a certain model/graph.
2. The coefficient of determination is the ratio of the explained variation to the total variation.
3. The coefficient of determination is such that $(0 <= r^2 <= 1)$, and denotes the strength of the linear association between X and Y.
4. The coefficient of determination represents the percent of the data that is the closest to the line of best fit. For example, if r = 0.922, then $r^2 = 0.850$, which means that 85% of the total variation in Y can be explained by the linear relationship between X and Y (as described by the regression equation). The other 15% of the total variation in Y remains unexplained.
5. The coefficient of determination is a measure of how well the regression line represents the data. If the regression line passes exactly through every point on the scatter plot, it would explain all of the variation. The further the line is away from the points, the less it is able to explain.

8.5.3 Statistical Formulas for the Mean, Variance, and Standard Deviation

The arithmetic mean of a set of data is denoted by Ybar; it is simply the arithmetic average of all the observations, That is,

$$\text{Ybar} = (Y1 + Y2 + \cdots + Yn)/n$$

The variance is a numerical value used to indicate how widely data in a group vary. Individual observations vary greatly from the group mean; the variance is bid; and vice versa.

The variance of a sample is denoted by the following formula. That is,

$$S^2 = (y1 - \text{Ybar})^2 + (Y2 + \text{Ybar}^2) + \cdots + (Yn - \text{Ybar})^2/n - 1$$

Standard deviation is a measure of how spread out the numbers are; is the square root of the variance. That is,

$$S = \left[(y1 - \text{Ybar})^2 + Y2 + \text{Ybar}^2 \right) + \cdots + (Yn - \text{Ybar})^2/n - 1\frac{1}{2}$$

Table 8.2 Illustration of Computation of Standard Deviation for Large Bore Nominal Pipe Size

Observations	Y1	Y1−Ybar	(Y1−Ybar)²
1	0.18	−0.63	0.39
2	0.21	−0.59	0.35
3	0.28	−0.52	0.27
4	0.42	−0.38	0.15
5	0.56	−0.24	0.06
6	0.70	−0.10	0.01
7	0.84	0.04	0.00
8	0.98	0.18	0.03
9	1.12	0.32	0.10
10	1.26	0.46	0.21
11	1.40	0.60	0.36
12	1.68	0.88	0.77
Ybar =	**0.80**	**0.00**	**2.70**
Variance	S^2 = **0.25**		
Standard Deviation	S = **0.50**		

8.5.4 Analysis of Historical Man Hours for Field Erection of Triple Pressure w/Reheat HRSG

The following table is based on actual direct craft man hours from job cost by cost code and type for field erection of heat recovery steam generators.

Table 8.3 Historical Man Hour Data for Erection of Heat Recovery Steam Generators (HRSG). Combined Cycle Power Plants: Field Erection of Similar Units Built in Successive Order

2.12 HRSG Triple Pressure; Three Wide-Installation Man Hours

Projects	1	2	3	4	5	Total	Ybar	Estimate
Average for No. of Units	4	2	2	2	2	MH	Unit MH	MH
Task Description								
HRSG—module casing	4527	5900	5328	5608	7500	28,863	5773	6528
HRSG—duct work SCR and inlet	2032	1815	1941	3723	1275	10,786	2157	3606
HRSG—erect gas baffles	1774	2777	2830	2156	896	10,433	2087	2578
HRSG—modules	840	1468	587	829	2960	6684	1337	1111
HRSG—pressure vessels	468	410	325	297	142	1642	328	421
HRSG—install duct burner			1927	539		2466	1233	1160
HRSG—platforms	2790	6575	4000	3289	6690	23,344	4669	4300
HRSG—skids	422	850	1608	1342	346	4568	914	240
HRSG—stack and breeching	2214	3230	803	3079	3356	12,682	2536	2898
HRSG—SCR and CO_2 internals	990	1138	927	2387	786	6228	1246	1417
Total man hours	16,057	24,163	20,276	23,248	23,949	107,693	22,278	24,258

8.5.5   Illustrative Example Unit Quantity Method for HRSG Module Installation

Scope of Field Work for Installation of HRSG Modules

Installation of Modules
All modules shipped with roof casing shop installed
All roof casing penetrations/packing glands shop installed
Unload modules
Set up lifting sled for each module
Lift and install modules

Unit Quantity Method

The method starts with the quantity take-off arranged in the erection sequence required to assemble and install the equipment. The estimator selects the task description by defining the work scope for the item to be installed. Each task is related to and performed by direct craft and divided into one or more subsystems that are further divided into assemblies made up of construction line items.

Table 8.4 **Task Description—Field Scope of Work HRSG Module Installation**

Estimate - Modules		Historical			Estimate		
Description	MH	Quantity	Unit	Quantity	Unit	BM
HRSG—Modules						**1111**
Reheater #3 HP superheated #2/ reheater #2	2.20	56.0	ton	56.0	ton	123
HP superheated #1/reheater #1	1.80	65.5	ton	65.5	ton	118
HP evaporator/IP superheated #2	1.40	165.5	ton	165.5	ton	232
HP economizer #2/IP super-heated #1	1.50	147.0	ton	147.0	ton	221
HP economizer #1/IP econo-mizer #1/LP evaporator	1.50	138.5	ton	138.5	ton	208
Feed water heater #2	2.20	55.0	ton	55.0	ton	121
Feed water heater #1	2.50	35.5	ton	35.5	ton	89

8.5.6   Fitting Straight Line by Regression—HRSG Module Installation

X	Y
2.20	56.0
1.80	65.5
1.40	165.5

Continued

X	Y
1.50	147.0
1.50	138.5
2.20	55.0
2.50	35.5

Regression Equation

Calculations

$$n = 7.00$$
$$\text{Xbar} = \mathbf{1.87}$$
$$\text{Ybar} = \mathbf{94.71}$$
$$\text{Intercept } a = \text{Ybar} - b\,(\text{Xbar})\ 312.98$$
$$\sum XY = 1110.80$$
$$n\,(\text{Xbar})\,(\text{Ybar}) = 1240.76$$
$$\sum X^2 = 25.63$$
$$n(\text{Xbar})^2 = 24.52$$
$$\sum X_i Y_j - n\,(\text{Xbar})\,(\text{Ybar}) / \sum X_i^2 - n(\text{Xbar})^2 - 116.63$$
$$Y = a + bX = -116.63X + 312.98$$

CorrelationCoefficient : $R = n\left(\sum XY\right) - \left(\sum X\right)\left(\sum Y\right)/n\left(\sum X^2 - \left(\sum X\right)^2\right)$
$$\tfrac{1}{2}\left(n\left(\sum y^2 - \left(\sum Y\right)^2\right)\right)\ \tfrac{1}{2} = 0.9415$$
CoefficientofDetermination : $R2 = (R)^2 = 0.8865$

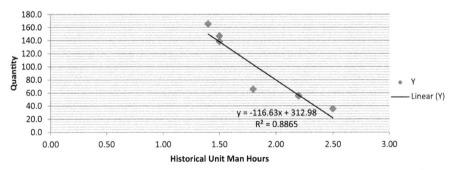

Figure 8.2 Regression line of $Y = -116.63X + 312.98$ showing deviations from regressed values.

The coefficient of determination is $R^2 = 0.8865$ and the correlation coefficient $R = 0.9415$ is a strong indication of correlation. 88.6% of the total variation on Y can be explained by the linear relationship between X and Y (described by the regression equation; $Y = -116.63X + 312.98$). The relationship between X and Y variables is such that as X increases, Y also increases.

8.5.7 Application of the Learning Curve in Construction

When a task of work is repeated without interruption, by experienced craft, the repetitive task requires less time and effort. Estimators in construction can apply the theory to productivity and future bidding of similar work. The "learning curve" principle can be applied to industrial construction and is useful for estimating: work that is comparable but work scope quantities may significantly differ and for forecasting output, time and man hours. The power equation models the "learning" concept and is used to estimate construction projects. If we use the comparison method for repeat projects, then it takes less time (man hours) to erect what has been previously erected, which is the learning process. The principle can be applied to industrial construction and modeled by using the "U learning model," power equation and regression analysis.

U Learning Model

Learning curves are mathematical models used to estimate efficiencies gained when an activity is repeated. The use of learning curves is to estimate the labor hours in construction when the scope of work is repeated. Learning effects are greatest when the erection process is manual.

Unit (U) Model

Used for comparing specific units of production and to fit a U curve to historical data.

The U model is based on the "power law" exponential equation, $y = ax^b$, where a and b are constants. If $b = 1$ the equation is a straight line passing through the origin, with a slope a.

Define the U model: $Hn = H1\ (n^b)$

where Hn = hours required for the nth unit of production; $H1$ = hours required for the first unit; b = natural slope of learning curve, illustrating if learning is rapid or slow If H1 and b are known, and the unit of production, the estimate of the hours for unit n can be calculated.

Example 1: Historical data collected from first unit on project 1 is 19,230h, assume $b = -0.15$, how many hours will the fourth unit require?

$$H4 = H1\ \left(n^b\right) = (19230) \times (4)^{(-0.15)} = 15619.6$$

System for Measuring Slope

The system for measuring slope, i.e., rate of learning, measures rate of learning on a scale of 0–100, in percentage.

Slope of 100% = no learning and 0% = infinity rate of learning.

In practice, effective range of industrial learning is 70%–100%.

Natural slope b is defined by the formula: $S = 10^b \log(2) + 2$ logarithm to base 10.

Example: if $S = 100\%$ (no learning)

$$b = \log\ (100/100)\ /\log\ (2)\ = 0$$

When substituting $b = 0$ into $Hn = H1$, n^b results in $Hn = H1$ for any value of n (no learning).

Example: if $b = -0.2$ value of S?

$$S = 10^{(-0.2)} \log\ (2)\ + 2 = 87\ \%$$

8.5.8   Prediction for the Total Hours for a "Block" of Production

Define man hours for a block of erection as the total man hours required to erect all units from unit M to another unit N, N > M.

TM, N is defined as follows:

$$TM,\ N = H1\ \left[M^b + (M+1)^b + (M+2)^b + \cdots + N^b \right]$$

Approximation Formula:

$$TM,\ N = [H1/(1+b)]\ [(N+0.5)^{(1+b)} - (M-0.5)^{(1+b)}]$$

Example 2: Historical data for the erection of four HRSG units at project 1.

Unit	Man Hour
1	19,230
2	17,076
3	15,776
4	15,226
Total project man hours	**67,308**
Ybar	16,827

Analysis of the historical data, for HRSG erection at project 1, unit 1 spent 19,230 man hours. Estimate the total man hours to erect the four units, units 1–4, if the learning curve has a slope of 85%.

$$TM, N = [H1/(1+b)][(N+0.5)^{(1+b)} - (M-0.5)^{(1+b)}]$$

Value for b: $b = \log (S/100)/\log (2)$ where S is slope covert to decimal
$b = \log (85/100) \log (2) = \log (0.85) \log (2) = -0.02125$
then, $1 + b = 1 - 0.02125 = 0.9787$

Compute H1 using $Hn = Hm (n/m)^{b}$; $H1 = (19230)(1/4)^{(-0.02125)} = 19805$
Calculate: $TM, N = [H1/(1+b)][(N+0.5)^{(1+b)} - (M-0.5)^{(1+b)}]$; with $N = 4$ and $M = 1$
$TM, N = (20236)[(4.5^{(0.9508)}) - ((0.5^{0.9508}))] = (20236)(3.8507)$
$= 77923$ Man hours

Actual historical man hours required to install the four units is 67,308 that results in a 8.6% saving due to learning.

8.5.9   Linear Regression—Fitting U Model to Unit Historical Data

$$y = ax^{b}$$

where y = hours required for the nth unit of production; a = hours required for the first unit; b = natural slope.

The power function $y = ax^b$ is transformed from a curved line on arithmetic scales to a straight line on log-log scales, let

$y = \log y$
$a = \log a$
$x = \log x$

taking logarithms of both sides, $\log y = \log a + x \log b$, appears like, $y = a + bx$.

Table 8.5 Regression Analysis of Illustrative Man Hour Data to Find Intercept a and slope b

Given $y = ax^b$ transformed to, $\log y = \log a + b \log x$

which is of the form, $y = a + bx$

Let $y = \log y$, $a = \log a$ Intercept, $b = b$ slope, and $n =$ sample size

Linear regression is a method for fitting linear equations of the form $y = a + bx$ to a set of x, y data pairs.

Example 3: Historical data for project 1; four HRSG's have been erected with no loss of learning between units. Fit U model to the following historical data: hours $= Y/100$.

Units (X)	Hours (Y)
1	192.3
2	170.8
3	157.8
4	152.3

Unit	MH	$X = \log X$	$Y = \log Y$	$(\log X)^2$	Log X Log Y
X	Y				
1	192.3	0.0000	2.2840	0.0000	0.0000
2	170.8	0.3010	2.2324	0.0906	0.6720
3	157.8	0.4771	2.1980	0.2276	1.0487
4	152.3	0.6021	2.1826	0.3625	1.3140
		1.3802	**8.8969**	**0.6807**	**3.0348**

$$\frac{\sum (\log X)^2 \sum (\log Y) - \sum (\log X) \sum (\log X)(\log Y)}{/n \sum (\log X)^2 - \left(\sum \log X \right)^2}$$

$\log a = 2.2835$

$$\frac{n \sum (\log X)(\log Y) - \sum (\log X) \sum (\log Y)}{/n \sum (\log X)^2 - \left(\sum \log X \right)^2}$$

$\log b = -0.17185$

	Then	
	$\log Y =$	$\log a + b \log x$
	$\log Y =$	$2.2835 - 0.1719 X$ (in logarithm units)

$X = \log X$	$Y = \log Y$
0.0000	2.2840
0.3010	2.2324
0.4771	2.1980
0.6021	2.1826

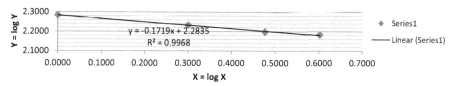

Figure 8.3 Regression line of Y = –0.1719X + 2.2835 showing deviations from regressed values.

CorrelationCoefficient: $R = n\left(\sum XY\right) - \left(\sum X\right)\left(\sum Y\right) \Big/ \left(n\left(\sum X^2\right) - \left(\sum X\right)^2\right)$

$$\tfrac{1}{2} - \left(n\left(\sum Y^2\right) - \left(\sum Y^2\right)\right)\tfrac{1}{2} = 0.9984$$

Coefficient of Determination : $R2 = (R)^2 = 0.9968$

The coefficient of determination is $R^2 = 0.9968$ and the correlation coefficient $R = 0.9984$ is a strong indication of correlation. 99.68% of the total variation on Y can be explained by the linear relationship between X and Y (described by the regression equation; $Y = -0.1719X + 2.2835$).

The relationship between X and Y variables is such that as X increases, Y also increases.

Appendix A: Formulas—Areas and Volumes

Square: Area = $(edge)^2$; $A = a^2$

Rectangle: Base × altitude; $A = ba$

Right triangle: Area = 1/2 base × altitude; $A = 1/2\,ba$

Pythagorean theorem:

$(Hypotenuse)^2$ + sum of squares of two legs of right triangle

$c^2 = a^2 + b^2$; $a = (c^2 - b^2)^{1/2}$

Oblique triangle: Area = 1/2 base × altitude; $A = 1/2\,bh$;

$A = [s(s - a)(s - b)(s - c)]^{1/2}$; where $s = (a + b + c)/2$

Parallelogram: Opposite sides are parallel; $A = bh$

Trapezoid: One pair of opposite sides parallel;

$A = 1/2$ sum of bases × altitude; $A = 1/2\,(a + b)\,h$

Circle: Circumference = 2 (pi) (radius) = (pi) (diameter); $C = 2(pi)R = (pi)D$

Area = (pi) $(radius)^2$ = (pi/4) $(diameter)^2$: $A = (pi)R^2 = (pi/4)D^2$

Sector of circle: Area = 1/2 radius × arc; $A = 1/2\,Rc = 1/2\,R^2$ angle

Segment of circle: Area (segment) = Area (sector) − Area (triangle);

$A = Rc - 1/2\,ba$

Ellipse: Area = (pi)ab

Parabolic segment: Area = 2/3 ld

Right circular cone: $V = (pi)r^2h$; A = side area + base area;

$A = (pi)r[r + (r^2 + h^2)^{1/2}]$

Right circular cylinder: $V = (pi)r^2h = (pi)d^2h/4$;

A = side area + end areas = 2 (pi)r (h + r)

Appendix B: Metric/Standard Conversions

Lengths
Metric Conversion
    1 centimeter = 10 millimeters; 1 cm = 10 mm
    1 meter = 100 centimeters; 1 m = 100 cm
Standard Conversions
    1 foot = 12 inches; 1 ft = 12 in.
    1 yard = 3 feet; 1 yd = 3 ft
    1 yard = 36 inches; 1 yd = 36 in.
Metric-Standard Conversions
    1 millimeter = 0.03937 inches; 1 mm = 0.03937 in.
    1 centimeter = 0.39370 inches; 1 cm = 0.39370 in.
    1 meter = 39.39008 inches; 1 m = 39.37008 in.
    1 meter = 3.28084 feet; 1 m = 3.28084 ft
    1 meter = 1.0936133 yards; 1 m = 1.09936133 yd
Standard-Metric Conversions
    1 inch = 2.54 centimeters; 1 in. = 2.54 cm
    1 foot = 30.48 centimeters; 1 ft = 30.48 cm
    1 yard = 91.44 centimeters; 1 yd = 91.44 cm
    1 yard = 0.9144 meters; 1 yd = 0.3144 m

Volumes
Metric Conversion
    1 cubic centimeter = 1000 cubic millimeters; 1 cu cm = 1000 cu mm
    1 cubic meter = 1 million cubic centimeters; 1 cu m = 1,000,000 cu cm
Standard Conversions
    1 cubic foot = 1728 cubic inches; 1 cu ft = 1728 cu in.
    1 cubic yard = 46,656 cubic inches; 1 cu yd = 46,656 cu in.
    1 cubic yard = 27 cubic feet; 1 cu yd = 27 cu ft
Metric-Standard Conversions
    1 cubic centimeter = 0.06102 cubic inches; 1 cu cm = 0.06102 cu in.
    1 cubic meter = 35.31467 cubic feet; 1 cu m = 35.31467 cu ft
    1 cubic meter = 1.30795 cubic yards; 1 cu m = 1.30795 cu yd
Standard-Metric Conversions
    1 cubic inch = 16.38706 cubic centimeters; 1 cu in. = 16.38706 cu cm
    1 cubic foot = 0.02832 cubic meters; 1 cu ft = 0.02832 cu m
    1 cubic yard = 0.76455 cubic meters; 1 cu yd = 0.76455 cu m

Areas
Metric Conversion
    1 sq centimeter = 100 sq millimeters; 1 sq cm = 100 sq mm
    1 sq meter = 10,000 sq centimeters; 1 sq m = 10,000 sq cm

Standard Conversions
   1 sq foot = 144 sq inches; 1 sq ft = 144 sq in.
   1 sq yard = 9 sq feet; 1 sq yd = 9 sq ft
Metric-Standard Conversions
   1 sq centimeter = 0.15500 sq inches; 1 sq cm = 0.15500 sq in.
   1 sq meter = 10.76391 sq feet; 1 sq m = 10.76391 sq ft
   1 sq meter = 1.19599 sq yards; 1 sq m = 1.9599 sq yd
Standard-Metric Conversions
   1 sq inch = 6.4516 sq centimeters; 1 sq in. = 6.4516 sq cm
   1 sq foot = 929.0304 sq centimeters; 1 sq ft = 929.0304 sq cm
   1 sq foot = 0.09290 sq meters; 1 sq ft = 0.09290 sq m
   1 sq yard = 0.83613 sq meters; 1 sq yd = 0.83613 sq m

Glossary of Key Terms

Definitions for Productivity Analysis

Actual time The time reported for work, which includes delays, idle time, and inefficiency, as well as efficient effort.

Allowance An adjustment for work, which includes delays, idle time, and inefficiency, as well as efficient effort.

Continuous timing A method of time study where the total elapsed time from the start is recorded at the of each element.

Cycle The total time of elements from start to finish.

Delay allowance One part of the allowance included in the standard time for interruptions or delays beyond the worker's ability to prevent.

Element A subpart of an operation or task separated for timing and analysis; beginning and ending points are described, and the element is the smallest part of an operation observed by time study. The length of the element can vary from minutes to hours to days, etc.

Fatigue allowance An allowance based on physiological reduction in ability to do work, sometimes included in the standard time.

Idle time An interval in which the worker, equipment, or both are not performing useful work.

Normal time An element or operation time found by multiplying the average time observed for one or multiple cycles by a rating factor.

Observed time The time observed on the stopwatch/electronic clock or other media and recorded on the time study sheet or media tape during the measurement process.

Person/man hour A unit of measure representing one person working for 1 h.

Productivity The amount of work performed in a given period. It is usually measured in units of work per man hour, man month, or man year.

Productivity rate A unit rate of production; the total amount produced in a given period divided by the number of hours, months, or years.

Rating factor A means of comparing the performance of the worker under observation by using experience or other benchmarks; additionally, a numerical factor is noted for the elements or cycle; 100% is normal, and rating factors less than or greater than normal indicate slower or faster performance.

Standard time Sums of rated elements that have been increased for allowances.

Task/operation Designated and described work sublet to work measurement, estimating, and reporting.

Piping Abbreviations

BOT Bottom
BW Butt weld
C to C Center to center

COL Column
CONN Connection
CS Carbon steel

CV Control valve
DIA Diameter
EL Elevation
ELO Elbolet
FAB Fabrication
FF Flate face
FLG Flange
FT Feet
FW Field weld
ID Inside diameter
IN Inch
ISO Isometric
LB Large bore pipe
LG Long
LOL Latrolet
MATL Material
No Number
NOM Nominal
OD Outside diameter
Olet Branch connection
PL Plate
PS Pipe support
PWHT Postweld heat treatment

RF Raised face
SB Small bore pipe
SCD Screwed
SCH Schedule
SMLS Seamless
SOL Slip on
SOL Sockolet
SS Stainless steel
STD Standard
STM Steam
STR Straight
SUPPT Support
SW Socket weld
TOL Threadolet
TOS Top of steel or support
TYP Typical
VA Valve
VERT Vertical
WN Weld neck
WOL Weldolet
WT Wall thickness
XS Extra strong
XXs Double extra strong

Labor Factors

Dilution of supervision This occurs when supervision is diverted from productive, planned, and scheduled work to analyze and plan contract changes, expedite delayed material, manage added crews, or other changes not in the original work scope and schedule. Dilution is also caused by an increase in manpower, work area, or project size without an increase in supervision.

Errors and omissions Increases in errors and omissions impact on labor productivity because changes are then usually performed on a crash basis, out of sequence, cause dilution of supervision, or any other negative impacts.

Fatigue Fatigue can be caused by prolonged or unusual exertion.

Joint occupancy This occurs when work is scheduled utilizing the same facility or work area that must be shared or occupied by more than one craft, and not anticipated in the original bid or plan.

Learning curve When crew turnover causes new workers to be added to a crew or additional manpower is needed within a crew, a period of orientation occurs to become familiar with changed conditions. They must then learn work scope, tool locations, work procedures, and so on.

Morale and attitude Spirit of workers based on willingness, confidence, discipline, and cheerfulness to perform work or task can be lowered due to a variety of issues, including increased conflicts, disputes, excessive hazards, overtime, overinspection, multiple contract changes, disruption of work, rhythm, poor site conditions, absenteeism, unkempt work space, and so on.

Overtime Scheduling of extended work days or weeks exceeding a standard 8-h work day or 40-h work week lowers work output and efficiency, physical fatigue, and poor mental attitude.

Reassignment of manpower When workers are reassigned, they experience unexpected or excessive changes, losses caused by move-on or move-off, reorientation, and other issues that result in a loss of productivity.

Ripple This is caused when changes in other trades' work then affects other work, such as the alteration of schedule.

Stacking of trades This occurs when operations take place within physically limited space with other contractors resulting in congestion of personnel, inability to use or locate tools conveniently, increased loss of tools, additional safety hazards, increased visitors, and prevention of crew size optimum.

Weather/season changes Performing work in a change of season. Temperature zone, or climate change resulting in work performed in either very hot or very cold weather, rain or snow, or changes in temperature or climate that can impact workers beyond normal conditions.

Index

'*Note*: Page numbers followed by "f" indicate figures and "t" indicate tables.'

Printed in the United States
By Bookmasters